# TIBETAN MONKS ANTI GRAVITY METHOD

# NEW THEORY FOR ANTI GRAVITY APPLICATION

**UNITED STATES COPYRIGHT OFFICE**
PAPERBACK
ISBN: 9798322713906
First Edition
16 April 2024
Johnny.vincento@yahoo.com
**ALL RIGHTS RESERVED**

**MAY GOD SAVE SOME OF MY BOOKS AFTER THE END TIMES TO GUIDE THE PEOPLE OF EARTH FOR NEXT 7000 YEARS**

# *INTRODUCTION*

*As the leading Applied physicist in full person Teleportation, this research can bring new knowledge to understanding anti gravity application and provide a working knowledge of Space explorations for our galaxy and others. No, not a sci fi fantasy, for real! In fact, Give me 10 disciplined students and we can advance humanities understanding of existence by 12,000 years. For the farther you go back in time, the more advanced. Yes, there were cavemen and all of that, but there were super civilizations that gained their knowledge from what Thoth said "From the Star born Races."*

*This research brings you the notes that found a world stage "THE ANTI GRAVITY PAPERS." These are not top secret, for how can they be if I am the inventor? Individual countries would like to have these papers for themselves to have this technology. So they made their way to the public domain to avoid that. No patents, no ownership. All are welcome to copy the 12 original papers of notes and sell,* *profit or patent them yourselves. You must include **"By Physicist Johnny Vincento."** Although, they are just the basics on how to achieve anti gravity and a gravity engine, it is a huge advancement for travelling space and possibly the universe: As well as flying cars and anything else one can think of.*

*Just to be clear, the information of this new frontier doesn't exist, so I will have to create some illustrations and examples so all can understand the workings of these amazing new additions to the human race.*

*The Tibetan Monks have already achieved this, documented in 1939 by Swedish engineers who witnessed this amazing feat of physics and engineering. With detailed descriptions their words and drawings are presented in this work. Testing can start now. Within a short few years humanity will have anti gravity applied to an endless area of avenues. This research gives a new real interpretation on what outer space is, explaining Dark Matter and gravity in a new undersstanding.*

*Before the discussions on anti gravity and the new theory of creating a gravity engine, a whole new theory to replace the Big Bang is included. The world won't comprehend otherwise, due to the present day understanding of physics. It is outdated and at a complete stop for advancing the human race. This will bring the world up to date on what the TELEPORTATION PROJECT discovered.*

# **TABLE OF CONTENTS:**
### *1. PEER REVIEW: NEW THEORY TO REPLACE THE BIG BANG*

### *2. 3D UNIVERSAL DARK MATTER FIELD - FABRIC OF SPACE EXPLAINED - DIMENSION SEPARATION VIEWER*

### *3. "THE ANTI GRAVITY PAPERS"*

### *4. TIBETAN MONKS LIFT STONES WITH ANTI GRAVITY SOUND FREQUENCY.*

### *5. Did ancients have fantastic technology to open dimensions with staffs that blast vibration waves?*

### *6. THE RETURN OF MAITREYA - STOPPING THE WHEEL OF SAMSARA - THE END OF THE WORLD AND THE BEGINNING OF THE SATYA YUGA 2030AD.*

TIBETAN MONKS ANTI GRAVITY METHOD

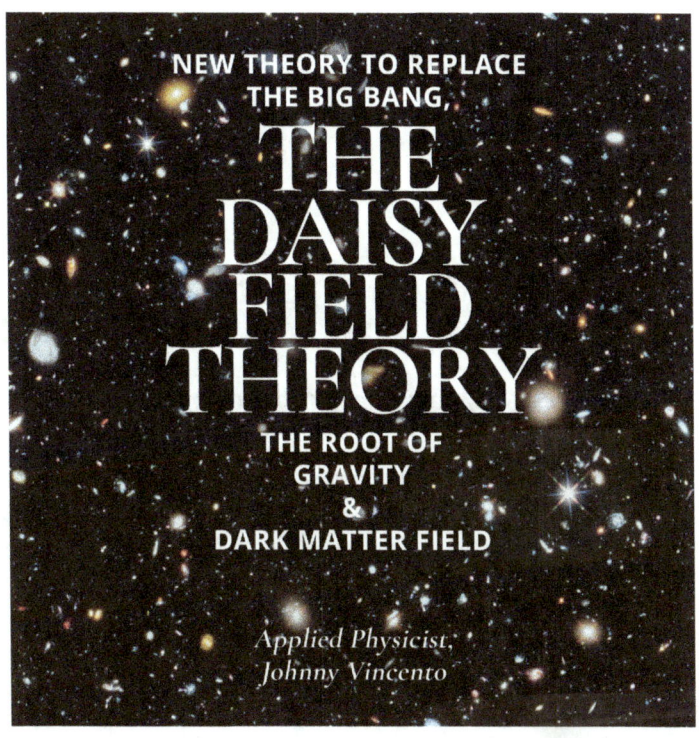

**UNITED STATES COPYRIGHT OFFICE**

**FIRST EDITION 2005**

**SECOND EDITION 2020**

**Published**

**February 21, 2022**

**Library of Congress**

**ALL HUBBLE PHOTOS ARE FROM NASA**

*THE ONLY WAY TO EXPLAIN
GRAVITY IS TO UNDERSTAND*

## THE DARK MATTER FIELD

## AND

## THE CREATION OF MATTER,

## WHICH IS THE ABSOLUTE ROOT OF GRAVITY.

From: Applied Physicist Johnny Vincento

SUBMISSION DATE: 01/08/2022

**"WRITTEN FOR THE STUDENTS OF THE WORLD"**

## ABSTRACT:
# Student Peer Review

*A new exciting discovery will bring great knowledge to humanity concerning the root of gravity and the universe itself. "The Dark Matter Theory" "The Daisy Field Theory" and "The Root of Gravity."*

### *DARK MATTER THEORY*

*Dark matter is described as a form of matter thought to account for about 85% of the matter in the universe. Its presence is implied in a variety of observations. This is a wrong conclusion, for it has been found that Dark matter is not matter.*

### *This next image is:* **CIRCINUS CLOUD IRAS 145686304.**

*(Here we have a NASA image from the Hubble Telescope.*

*This Whitehole is blasting out matter into our universe from an overlapping universe. It also seems to be also* **blasting out Dark Matter itself,** *along with matter into our universe.*

**This would explain our expanding universe.**

*Notice that you can not even see a star behind this possible Dark Matter being ejected. Dark Matter gravitational effects cannot be explained*

*by accepted theories of gravity, because no one until this Peer Review, knows what exactly Dark Matter is. It does not absorb, reflect or emit electromagnetic radiation and is therefore difficult to detect. The following discussion will shed light on this newest theory. The boundary between universes and* *worlds seems to be a **VIBRATIONAL ENERGY FIELD** that is not as a wall. This field separates three dimensions somehow. Never the less, it can be opened.*

*This next NASA image, is of **Galaxy M82**. The thought again is that this is a Black hole, but as you can see, matter is blasting out, not going in.*

*WHITE AND BLACK HOLES are really just WORMHOLES or GIGANTIC UNCONTROLLED EINSTEIN ROSEN BRIDGES that naturally occur between dimensions and universes.*

*The closest overlapping universe has different laws of physics, which allow for particles to create matter randomly or **"UNOBSERVED"** into our universe. However, the Whitehole must stay open to allow for this effect. We do not have these natural laws of physics in our vibrational plane of existence unless a Wormhole is open.*

**WRITTEN ABOUT 12,500 YEARS AGO ON STONE TABLETS By ancient scientist Thoth.**

**"On earth, man is in bondage, bound by space and time to the earth plane. Encircling each planet, a wave of vibration, binds him to his plane of unfoldment."**

**"Know you man, that all space is filled by worlds within worlds, yes one within the other yet separated by law."**

"Once in a time long forgotten. I, Thoth, opened the doorway. Penetrated into other spaces and learned of the secrets concealed."

"Deep in the essence of Matter are many mysteries concealed. Nine are the interlocking dimensions and nine are the cycles of space."

"Nine are the levels of Consciousness and nine are the worlds within worlds. Space is filled with concealed ones. For space is divided by time. Seek you the key to the Time Space and you shall unlock the gate."

Know you that throughout the Time Space. Consciousness's surely exist. Though from our knowledge it is hidden. Yet it still forever exists."

"The key to worlds within thee are found only within. For man is the gateway of mystery and the key that is one with the one."

### *"You are the key to all wisdom. Within you is all Time and Space."*

"Listen you man, and hear a mystery stranger than all that lies beneath the sun."

"Yes in a time that is yet unborn,...Man, a perfect flame of this cosmos, shall move forward to a place in the stars. Yes, shall move even from out of this Space Time into another beyond the stars."

## *THE ROOT OF GRAVITY*

## *and*

## *THE 3D DARK MATTER FIELD*

*The root of Gravity is matter coming into our universe and bending Dark Matter. As said before, Dark Matter is not matter. It is just an energy field of vibration. They should actually rename it as **"The 3D Universal Field."** Picture a sink full of water and an orange ball is put halfway in. The water represents 3D Dark Matter field. Notice how the water bends with the Matter put into the field. The root of all matter and gravity in the universe is "THE DAISY FIELD THEORY."*

*Another way to visualize Dark Matter: Are tiny squares within squares **(or invisible RUBIX CUBES within RUBIX CUBES)** throughout all of the universe. Two layers overlapping: Right in front of you is Dark Matter. Yes, it has air in it, but it is still a 3D field separating dimensions.*

*The outer layer of RUBICS CUBES is the physical immortal world of Heaven, vibrating at a higher frequency of LIGHT and Matter. Energy is vibrating Matter.*

*The inner layer of RUBICS CUBES is our mortal Earth. The squares are all separated by a 3D field. So two... three dimensional worlds within each other. Separated by vibrational matter frequency. There are many other overlapping worlds, however, this is the closest Multiverse that opens to our Universe.*

*This is a new proven theory that takes in account of a closest overlapping universe occupying the same space, unseen due to dimensions. The proof is "The Teleportation Project."*

*This next NASA image is from Hubble showing the* **Galaxy Hercules A 3C348** *or* **Radio Galaxy** *some call it. This Whitehole (Supposed massive Black hole) is blasting out amazing jets of particles, matter and plasma from the hole (WORMHOLE STEM) itself. Many Galaxies can be seen shooting* *out matter light years long on each side. This is in fact a Whitehole and the reason everyone considers these Black holes, is due to humanity not knowing about an overlapping universe.*

*This galaxy is located:*

*About 2 billion light years from us and the galaxy in the center is about 1000 times bigger than our Milky Way.*

*These blast jets are calculated at 1.5 MILLION light years wide!!!*

**DARK MATTER DOES NOT CREATE GRAVITY THE WHITEHOLES BLASTING MATTER INTO OUR UNIVERSE BENDS THE DARK MATTER FIELD CREATING GRAVITY.**

**THIS IS THE ROOT OF ALL GRAVITY IN THE UNIVERSE.**

*The Daisy Field Theory, takes this research of a closest overlapping universe and challenges the belief that Hercules A 3C348 is a Black hole. This new theory says that this galaxy and every other galaxy shooting out matter, in the entire universe, has in fact Whiteholes (STEMS) at their centers. The Big Bang Theory needs to be updated or redone totally and The Daisy Field Theory is its replacement. From just looking at this image and the* **Deep Field images of Hubble (NEXT),** *all those lights are Galaxies.*

*In the next image of* **NASA Hubble Ultra Deep Field** *region of space, are about 10,000 lights. Here is a zoomed in image. Each light is an* *entire galaxy. They all seem to be developing at different stages.*

*The Daisy Field Theory says that there may not have been a Big Bang at all and each galaxy is SPROUTED LIKE A DAISY from an overlapping universe through a WHITEHOLE (STEM). Then after the SPROUTING of the*

*galaxy, it turns into a Black hole. Although, this theory only says that every galaxy was formed from a Whitehole. Throughout a galaxies life, the center hole can* **turn on and off from a Whitehole to a Black hole** *with no explanation as of yet.*

**This next NASA image,** *we have the* **M87 Galaxy** *blasting out matter about 5000 Light years long from a Whitehole. Again, everyone in the world thinks this is a Black hole doing this.*

*THE DAISY FIELD THEORY would explain a huge amount of Hubble images of galaxies. The image we have here (above) is a perfect example of new matter coming into our plane of existence...our universe.*

*Max Planck said, "All matter originates and exists only by virtue of a force which brings the particle of an atom to vibration and holds this most minute solar system of the atom together. We must assume behind this force the existence of a conscious and intelligent mind. This mind is the matrix of all matter."*

*Niels Bohr and others developed the* **"Copenhagen Interpretation,"** *stating that a quantum particle doesn't exist in one state or another as a particle or a wave, but in all of its possible states at once. When we observe its state, the particle is forced to choose one probability, and that's the state we observe. The particle may be forced into different observable states each time, which explains why a particle behaves erratically and can give differing results.*

*The **"Observer Effect"** states that the process of observing a particle changes the way the particle behaves.*

*Now add this "Observer effect" to the overlapping Earth and its universe with its different laws of physics that allow for the creation of matter by just visualizing and wishing for it: Or even speaking matter into existence.*

*While studying the Sanskrits and other ancient texts, there are accounts of some Monks visiting the overlapping world and making matter appear out of thin air. There is an overlapping universe and an overlapping Earth. (Actually many, but for this theory only one Multiverse and one Earth is needed for discussion).*

*I was able to create matter (A butter knife) out of thin air in the overlapping world by "Observing Matter." I visualized it and it appeared on the floor. I picked it up and I was amazed that it was solid stainless steel, just as I visualized it. The entire atmosphere seems to be the "String theory itself!" Invisible building blocks of creation allowing for a person to be Quantum Entangled, erasing death. It reminded me of Jesus being a doorway and creating Wine and bread and even fish by "Observing Matter." When the Wormhole is open, this can be done here on Earth apparently. In the Sanskrits, one story talks of a Monk creating matter (in the form of gold coins) by wishing for them.*

*Niels Bohr said to Einstein that "Matter is only created when **observed**." and "Einstein replied "Are you telling me that the Moon behind me isn't there, because I am not looking at it?"*

*Niels Bohr was doing his best to describe what I am saying. The overlapping universe has other laws of physics. One's mind is the transformer and one's life force creates matter from memory (Particles memory) which is how one can travel through dimensions in a Quantum Entangled exact duplicate of one's self as in the ancient Monks. We are made of particles and can be in two or more places at the same time physically. Even estimated at 12,500 years ago, stone tablets were found saying this:*

*"Nine are the worlds within worlds."*
*"Know that throughout the space that you dwell in are others as great as your own. Interlaced through the heart of Matter yet separated in space of their own."*

*That sounds like the Vibrational energy field that separates universes and worlds.*

*But what happens to matter that comes through that is **unobserved?** You have random matter creation from nothing. And that is exactly what is coming into our universe through Whiteholes. **Unobserved matter or (Particles) forming matter from nothing.** The Whitehole has the different laws of physics from that overlapping interlaced universe, which makes matter creation from nothing possible.*

*You seen the image of the CIRCINUS CLOUD taken by Hubble: IRAS 145686304 On each side of it West and East is a gigantic dark area around it. It is known as the Circinus Molecular Cloud. It blocks out the entire universe behind it. The star is blasting out material from its core. The cloud itself is some 2,280 light years away and is more than*

*180 light years across. The reason I brought this event up, is this may actually be Dark Matter coming into our universe along with random matter through a Whitehole.*

*The expanding universe theories, are that the Big Bang exploded and everything is moving away. That last part has been found to be true. Galaxies are moving away from us by proof of their light color. Does that mean the Big Bang is the only answer? The theory of a stretching universe like a rubber band is another one. Space bends all the time.*

**The name "Daisy Field Theory" came to mind when I seen the Hubble Ultra Deep Field images. Just as we have endless grass fields here on Earth full of Daisies, they are all sprouting on their own, individually, from their Whitehole stems and developing at different stages in each galaxies life time.**

**\*\*\* The Daisy Field Theory and The Root of Gravity says: That all GALAXIES, ALL MATTER in the universe is created by, SPROUTING FROM A WHITEHOLE from an overlapping universe and this is the ROOT OF GRAVITY BY BENDING THE DARK**

## *MATTER FIELD.* \*\*\*

*The key to going through the boundary is to equal the vibrational matter waves of the other higher frequency worlds field. This CAN be accomplished in a machine at least in theory. Dark Matter is in front of our faces.*

*Even Jesus said, "Know what is in front of your face, and what is hidden from you will be disclosed to you. For there is nothing hidden that will not be revealed.*

*There is just stuff in the air. It is a vibrational 3D energy field that when bent creates gravity.* **We can understand Dark Matter much better by going through it instead of trying to understand it with theories and equations on paper.**

JOHNNY VINCENTO

# The 3D universal Dark Matter Field.

The next discussion will explain how to go through it. This research is not about creating this imparticular portal. It is needed to understand the physics for creating a **gravity engine** and **anti gravity** itself.

*The Project proves, a 3D physical world overlapping ours, at a higher frequency of vibration matter existence.*

**The creation of a more stable communication device,** *would be final proof of Heaven's existence. The teleportation is a great subway transportation system for travelling the universes and parallel worlds, however, not many people are bold enough to want to*

*explore in this refound transit cosmic railway.*

***The creation of a more stable communication device,*** *between the closest overlapping would need technical engineering in the field of electrical and Vibrational Matter Waves (VMW's). This is a term that doesn't exist, even so, that is what needs to be discovered, tested and applied. The frequency of that higher frequency matter existence that occupies the exact same space right in front of us: That is the goal. To match those waves and open a portal to Heaven.*

*For the record, as recorded time is my witness: I will tell humanity, never attempt to contact any other world using lower frequency waves of vibration in this portal. The negativity will not make you feel well. If the higher world makes you feel ecstasy, it would be a good assumption to have the opposite effect contacting a lower vibration. Use Earth's frequency and go higher from there only. For example:*

*What is Earth's frequency?*

*The Earth acts as a huge electric circuit. Its electromagnetic field surrounds all things. That is a natural frequency pulsation of 7.83 hertz on average. It is called the "Schumann resonance," named after physicist Dr. Winfried Otto Schumann. Our brains are only operating at the Alpha medium frequency. We are only using half of our brain capabilities. Some people are using less, as you see in the world. The higher frequency LIGHT will raise = stimulate instantly increase our intellect to 100%. From 7.83 hertz to around 15.65 hertz would be an estimate. In issue #8 of TELEPORT NEWS, an investigation is looked into this LIGHT and how the ancients used it for immortality.*

*When a portal is made to Heaven, the LIGHT will shine on only the people in front of it. It is not like the movies where people can walk in and out. Those are different matter frequencies. Could your body walk in the world overlapping ours right now? Do you see people from Heaven walking around in our world? The degree of matter is at such a height of vibration, you would never even be able to take a step. A person could*

*sit in front of it and bathe in its LIGHT.*
***Interdimensional Communication*** *between worlds will also be established. Let's say hello to God.*

*If you think this is Sci-Fi nonsense, here is a portal opened in the Bible:*

## *THE ROAD TO DAMASCUS EVENT*

***Acts 9:3-9***
***King James Version***

*3 And as he journeyed, he came near Damascus: and* ***suddenly there shined round about him a light from heaven:***

*4 And he fell to the earth, and heard a voice saying unto him, Saul, Saul, why persecutest thou me?*

*5 And he said, Who art thou, Lord? And the Lord said, I am Jesus whom thou persecutest: it is hard for thee to kick against the pricks. (Means resisting and stabbing me)*

*6 And he trembling and astonished said, Lord, what wilt thou have me to do? And the Lord said unto him, Arise, and go into the city, and it shall be told thee what thou must do.*

*7 And the men which journeyed with him stood speechless, hearing a voice, but seeing no man.*

*8 And Saul arose from the earth; and when his eyes were opened, he saw no man: but they led him by the hand, and brought him into Damascus.*

*9 And **he was three days without sight**, and neither did eat nor drink.*

*Saul was a killer and after the exposure to the LIGHT he became an instant saint. He got baptized and wrote a large portion of the New Testament. So that is some proof of the effects of the LIGHT. It is at a greater frequency so that is what the particles do to our brain. We are dealing here in this subject with LIGHT from ANOTHER EXISTENCE. After I was exposed to the LIGHT when I explored Heaven, my brain has been changed and I can't do anything bad purposely and my whole consciousness was evolved in the sense that THE TRUTH was seen in the experience of existence within another realm or plane of reality greater than ours.*

## *THE D.S.V. IS ALREADY BUILT*

*As the leader in these new scientific discoveries, it is no wonder why someone from the past, present, or future, tried to contact me using what seems to be an actual D.S.V. of their own. This was proof that the machine can be built. Someone was checking on me from the world overlapping our world **or checking on me from across time.***

*In the fifth month of the year 2014 on the 16th day. This being a Friday. Sitting alone in my office chair with the lights off and the sky becoming dim. I was in distress, because of my present situation and happy too, thinking of what the voice meant. About two weeks before while putting on my "Mary with baby Jesus" necklace: A verbal voice in my right ear said "This is going to be the best year ever."*

*I am alone with my thoughts on that May 16th with my back to the wall in my desk chair. There is about a foot distance separating my chair from the wall.* **I hear above me! Electricity! "cheee cheee --- cheeeeeeee cheeeeeeee --- CHEE!"** *With this sound came a bright white "Light" that illuminated the room. Then with the last* **"CHEE!"** *sound the entire room was so bright that this type of "Light" doesn't exist on Earth. It was so bright that I was thinking "Don't look to where it is coming from." I thought I would get blind if looking directly at the "Light."*

*This opening came from about a foot and a half above my head: While I was sitting. The wall was behind me about a foot. So somewhere in that area. Out of the air or wall came hovering straight across the room an object that I will describe to you: It was bigger than a pinball machine ball. And smaller than a golf ball.*

*In between those two was the size. It looked like it was filled with liquid Mercury. At the same time had "Ruby blue" random spots like our Earth would look like if all the land was Ruby blue in color and the rest of the*

*planet was liquid Mercury. This didn't look anything like a mini Earth. I am describing the randomness of the color Ruby blue.*

*The device cruised very slow in a straight line about 18 feet and then disappeared.*

*I seen spirits and they look to be filled with liquid water. They are also larger. This it seems was a drone of some sort sent through dimensions to visit or view me. The substance may have had a seeing device inside. Why else would it bother to cross into our dimensions? That is a lot of effort to do such a thing. If it wasn't a lot of effort, things like that would happen all the time.*

*It was enclosed in something clear, (Plasma?) or a liquid ball full of Mercury and what else the Ruby blue was could be made by using a magnetic field? (As the ball?)*

*If it wasn't Mercury then it was silver metallic liquid paint. I doubt it was that. So that's why I am so sure it was Mercury or something that looked the same.*

*After it disappeared I felt the wall and it was not warm or hot or damaged in any way. No sign of it coming from the wall. So it had to have opened dimensions through the air itself.*

**Here is a real riddle that will be fun for Chemistry students to try and figure out. I don't have the answer. Only today, I had some time and focused on this contact from the future on our plane of existence or from Heaven itself.**

**Question one:** *Is there a clue to figure our frequency of their portal? This would be vibration waves.*

*Answer: Yes, The color royal blue scattered about on the liquid mercury within the transparent sphere.*

**Question two:** *What Chemistry metals turn that color blue?*

*Answer: Indium - Hydrogen - Mercury*

**Question three:** *What spectrum of light on the metal Indium would turn it the color royal blue.*

*As you can see in above chart, there is a specific wave frequency to obtain royal blue color for Indium. Looks to be between 410 - 451.14nm*

**Question four:** *Can the frequency of the portal that opened in my office be found converting nanometres to Hertz?*
Answer: Yes,
Convert wavelength in nanometres [nm] to hertz [Hz]
Frequency and Wavelength converter:
Here is very basic explanation to reach the royal blue color in Indium:

1 wavelength in nanometres = 2.99792458 x 10 + 17 hertz
451.14 wavelength in nanometers = 6.645220064725e+14 Hertz
A Nanometer (nm) is used to describe Ultraviolet light frequencies (wavelengths 380 to 10 nm).

A Hertz (Hz) is used to describe the frequency

*of sounds and mechanical vibrations (wavelengths 20 Hz to 20 kHz)*

*Here is how to reach the color blue frequency in Hydrogen*

*Here is how to reach the color blue in Mercury: What is the wavelength of mercury source?*

*The wavelengths of the prominent lines in mercury are purple (405 nm), blue (436 nm), blue- green (492 nm), greenish yellow (546 nm), yellow (577 nm), orange (623 nm), and red (691 nm)*

**Figure 4:** *Mercury spectrum in color blue is at 436 nm*

*We will examine two different light sources, mercury and hydrogen. In these sources a high voltage sent through the gas produces excitation of some of the atoms in the gas. When these atoms de-excite they emit radiation characteristic of the element, some of it in the optical wavelengths. For the most part, these de-excitations produce discrete spectral lines reflecting the quantized nature of energy levels in an atom.*

*As the portal opened I heard electricity sparking. It is possible that one element or all three were in the drone sphere: Indium and Hydrogen and Mercury?*

*Regardless, the Hertz mechanical vibration waves are:*

***Hydrogen turns blue at: 486 nm = 6.16856909465e+14 Hertz***

| Wavelength (nm) | Color | Frequency (Hz) |
|---|---|---|
| 435 | purple | $6.90 \times 10^{14}$ |
| 486 | blue | $6.17 \times 10^{14}$ |
| 657 | red | $4.57 \times 10^{14}$ |

*This being very interesting, for Hydrogen is the lightest element. Thus, a possible clue to how the ball drone floated. Maybe the Mercury didn't turn blue, but the Hydrogen. It is heavy, weighing 13.6 times more than equal volume of water. Maybe the Mercury surrounded the Hydrogen in the ball?*

***Indium turns blue at : 451.14 nm = 6.645220064725e+14 Hertz***

***Mercury turns blue at: 436 nm = 6.891780643678e+14 Hertz***

*\*\*\*photo is mercury turing blue at the frequency applied\*\*\**

*Note: Mercury is a metal that always stays in a liquid state unless it freezes into a soft solid like tin or lead at -38.83 °C (-37.89 °F).*

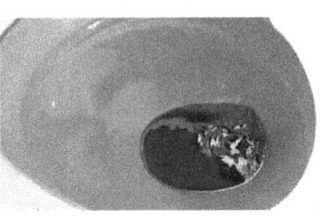

*This matches what I seen in the drone. So a starting point vibration frequency should be started at this hertz. The scientists may have used this visual tool within the drone to give the frequency to open dimensions. They would know that physicists are more in tune to these things. Plus chemistry was a born gift for reasons unknown.*

**The color blue being seen gives a clue that a frequency equal to the Hertz examples above is the frequency to use - to open a DSV MACHINE.** *If the scientists were from the future or anywhere on our plane of matter vibration, then that would be beneficial. However, if the scientists were from a different vibrational world, then what is the formula to convert? The Photon LIGHT particles were* *about 10 times higher than ours. Even so we now have a starting point.*

*Indium is a solid metal that can be melted: It looks like this:*

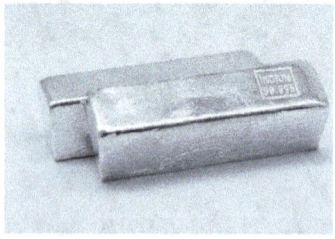

*Since no heat was felt on wall, Indium being a solid, most likely should not be considered part of the drone.*

*It was said by researchers that our universe and planet has the frequency of 7.4 Hertz. So these findings are well below this. There is a negative matter world as well. For there are worlds within worlds. Maybe that is the start. It doesn't make any sense.*

**Maybe you can solve this mystery my Chemistry & Engineering student friends.
Johnny.vincento@yahoo.com
Join the fun: I will put your solution in the book. You won't get paid. But your response will be worldwide.**

*One time a guy about 19, early twenties, Caucasian, clean shaven, thin face, wearing a cotton short rimmed foreign hat with the Sun flaps for the back of neck, wearing high top boots, with cargo cotton pants. small pockets on the legs, appeared out of thin air and just looked at me and disappeared. Was that one of my future students? He seemed like a nice guy just having fun Teleporting to see his Teacher. Or the one who gave the world Teleportation. His face was very thin from fasting I would guess.*

*Maybe, I have a fan group in the future? That would be nice.*

***Message to the scientists who contacted me. Your experiment worked amazing, wonderful, spectacular, perfect, in all ways. The LIGHT was so bright I couldn't even look at it!!! Very happy for your success. And apparently mine. Please make a Johnny Vincento Christian Science holiday in the future for me. Did the sphere drone you sent through have a recording device in it?***

*If so, please send another drone to communicate with me to show me how to use cameras through dimensions? Try to use a sound system and speak English please, either through the open portal, or through the drone itself. The portal only stayed open for about a minute. Please assist to add to my research to make it amazing by obtaining your knowledge in the future: Now in the year 2022. Use the coordinates for steering 20-14-05-16-28-08-North-H-Ar-Vey-A-V-* **For your successful Dimension Separation Viewer opening.** *You won't find me there now. Anytime after October 2022 you can try to contact me by just steering: using my name and year in the USA. Please also send information on creating anti gravity through the portal. If you left the DSV open for longer, You may have been able to speak through it?*

## *MY MISSION AS A SCIENTIST FOR THE D.S.V.*

*The D.S.V. can reboot the entire world into a planet of "LIGHT" by means of evolving "Homo Sapiens" into a "Child of Light." A "New Spiecis of advanced Humans of "Goodness" without darkness. We only use about 50% of our brain mass. The Light will activate the other that has been dormant for ages.. A natural "Heavenly LIGHT Transformation" reboot advancement of the human brain or intellect. When the "LIGHT" of the other world shines on you, it is very difficult to do bad things and that was just one dosage for me. This Heavenly Light is also the key to immortalily. I read the anciet writings from 12,500 years ago and I am a physicist in Quantum entanglement. Build the machine, the ancient scientist who lived over a thousand years wrote "100 years out of ever 1000 one must bathe in the Heavenly light to renew ones body." Sitting in front of it for 10% or on weekends of one's life isn't too much to do. Humans were created in the LIGHT!*

# *WHAT IS THE D.S.V. MACHINE?*

*The Dimension Separation Viewer is a window to lift the veil between our mortal physical world and the immediate closest physical world occupying the same space: Unseen due to dimensions and accessed successfully through Teleportation.*

*The Silver Cord Quantum Entanglement explorations prove without a doubt that other worlds overlap ours. Since that is the finding, that means each world then has its own entire frequency of matter existence throughout their universes, which makes the "Daisy Field Theory" a very valid conclusion. Not to mention the other laws of physics in those worlds, that are unfamiliar in our mortal plane.*

*All it will take is a team of engineers with a lab. The machine is as follows:*

*Setting up a round frame, with a diameter of six feet and one to three feet deep: The machines and equipment are attached to the frame. The air (Dark Matter 3D Vibrational Energy Field) within the frame is manipulated, to attempt to duplicate the natural occurrence of the atmosphere and to duplicate the "Road to Damascus Event." A great effort is made, to what has never been tried among man, A DIMENSION SEPARATION VIEWER!!!*

*Wheels should be on this unit, for this may not be a window but only a slice. Moving this around the room, one will attempt to see furniture and decor or even trees of the home or land that occupies the same space. When this occurs, the formula will be a success with all frequencies documented. The main goal however, is even more insane than one can imagine: To actually instantly evolve the human race into a new super human called the **"Novus human."***

*The LIGHT from the other world actually changes the brain or intellect of a human advancing that person into an evolutionary state instantly. After a few doses of this LIGHT that will come out of the D.S.V. window, a person will have been completely advanced into a new race of enlightened human. Let's look at why this machine can be built.*

The main application to make such a machine would be "Vibrational Waves." Furthermore, from my experience, building the "DSV" would also require wearing torch glasses or very heavy sunglasses. Light and matter are different across dimensions. When one Teleports, the physical body that is materialized and duplicated is immortal. The natural laws of physics during the Wormhole and Quantum Entanglement materialization to the second body: Aligns within that worlds "Vibrational Matter" and "Light." That is the reason that "Vibrational waves" are needed to see the different "Vibrational Matter." which occupies the same space.

Light is not the same and matter is not the same as here. The Teleportation is safe everytime, but when a "DSV" machine is built our bodies won't be aligned with that worlds light of different photon wave particles and frequencies. The photon particles of an overlapping worlds light are in tune with that worlds matter. The light from that world won't affect the person Teleporting due to being in a rematerialized physical body of that matter in tune to its worlds light. However, if a Wormhole was to open and it did on me once: I guess someone wanted to contact me with their own DSV. The LIGHT was so bright that if I would have looked at it I may have been blinded.

# "THE ANTI GRAVITY PAPERS"

TIBETAN MONKS ANTI GRAVITY METHOD

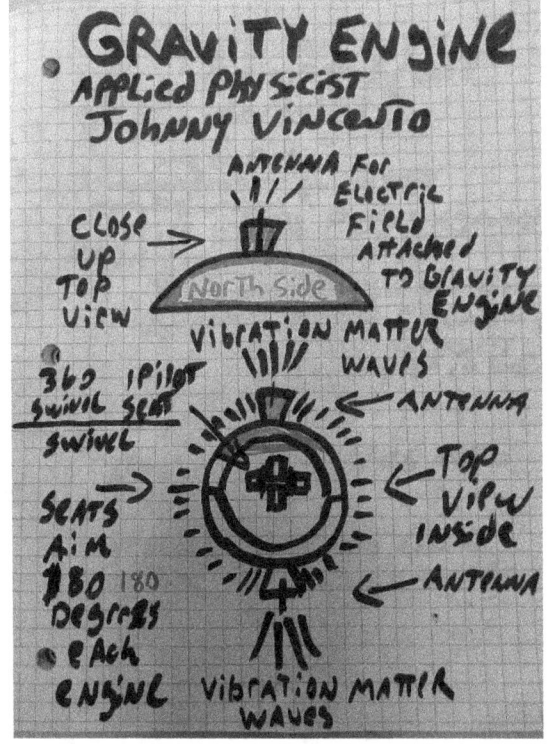

Both engines are attached on swivel. They will always be exact opposite directions. This will be for getting back to Earth. Once the frequency is found to open void in the Dark Matter field, it can be turned on to go forward. For example: Turn Gravity engine on for 5 seconds, then turn off. See how far spacecraft travelled. Then turn on opposite Gravity engine for 5 seconds to arrive where started.

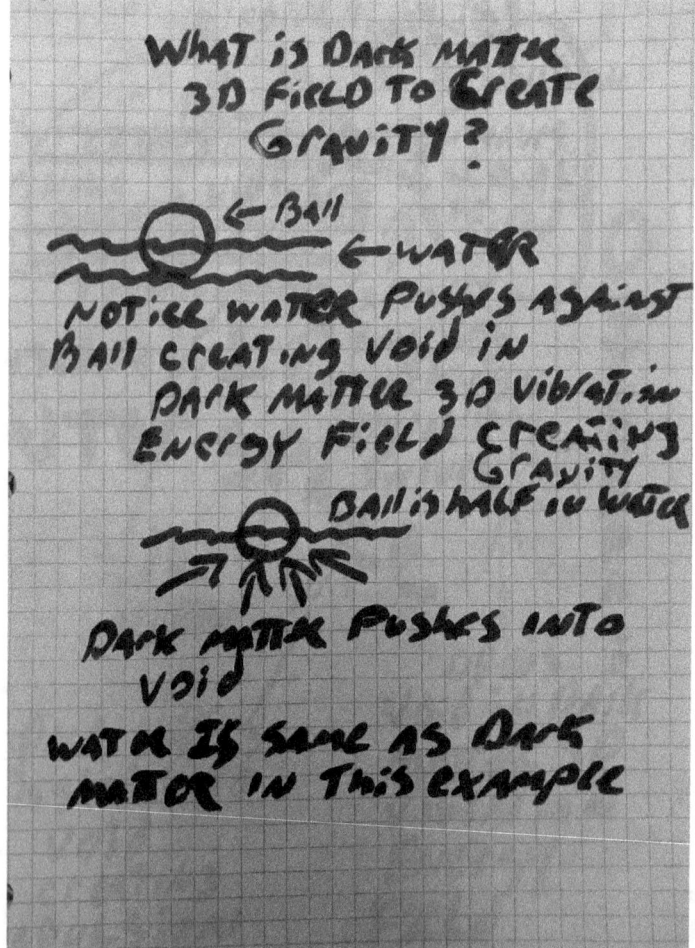

## TIBETAN MONKS ANTI GRAVITY METHOD

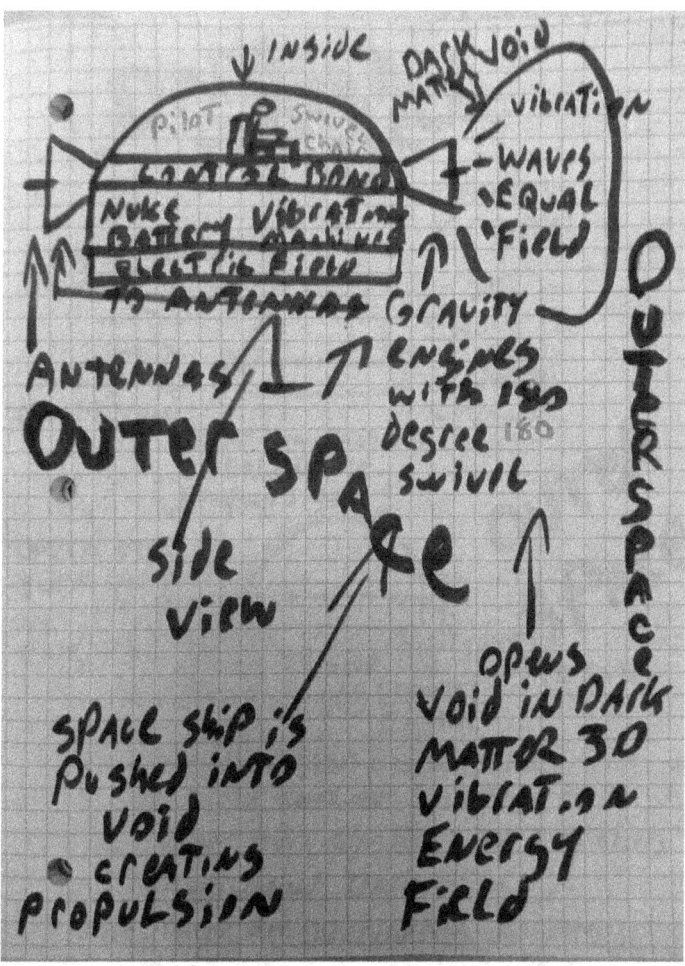

Many reports of flying saucers even from thousands of years ago describe simular aero dynamics, thus being still and speeding off in a straight line of light in a split second. That seems to be what this Gravity engine would accomplish, yet is unknown without testing. If any government wants me to work for them, I am more for civil engineering and UNIVERSE TRAVEL EXPLORATION AND SPACE TIME focus: To bring in benefits to humanity not war. Those past athiest scientists always sell out to money and war. What a waste, the scientist got a bunch of cash and millions or billions of people die or will die. I would never want my anti gravity ideas to turn into weapons. I will have no part in it. We care nothing about money. I don't care much about money. To me I am amazed how everyone is so greedy over it. It is just a tool to make all these people happy. So they can have their precious cash. I would rather just pay all my bills for the whole year just to be rid of the whole money thing. It is bothersome. Every single month everyone wants money. My possessions can fit in a bedroom closet besides some basic items such as furniture, tools etc...I don't even have a car. If that was the case and money was my God I would have contacted any one of these governments and said **"I'm world leading physicist Johnny Vincento, Do you want to obtain Time Travel or Teleportation which are already tested and working?"** Then I would be on their payroll under top secret non disclosure, with a million dollars or whatever amount I wanted per year. Yes, an Einstien Rosen bridge is available right now. To even contact aliens. To actually teleport right into their spacecraft. That is how precise the steering is of the other TELEPORTATION PROJECT. It hasn't been applied in that method, but other tests resulted in extreme precision steering all the way through space and time. I am a self taught Applied physicist, so being involved with these unapplied theories, they are all founded in the Applied findings of actual teleportation.

I will add that after the LIGHT of Heaven, or the closest overlapping world at the highest frequency of matter shined on me. My brain was raised a bit in function and frequency. I am no better than anyone, the higher LIGHT photons opened the unused portions of my brain capacity. To a next level up frequency I would calculate. Not enough to create a new species, however that wouldn't be difficult with building that other machine, THE D.S.V. This is my self diagnosis and a fact. Imagine if a person was able to use their entire brain by bathing in the LIGHT of Heaven. My brain frequency was only raised a small amount by being submerged in that LIGHT for a very short time.

# TIBETAN MONKS ANTI GRAVITY METHOD

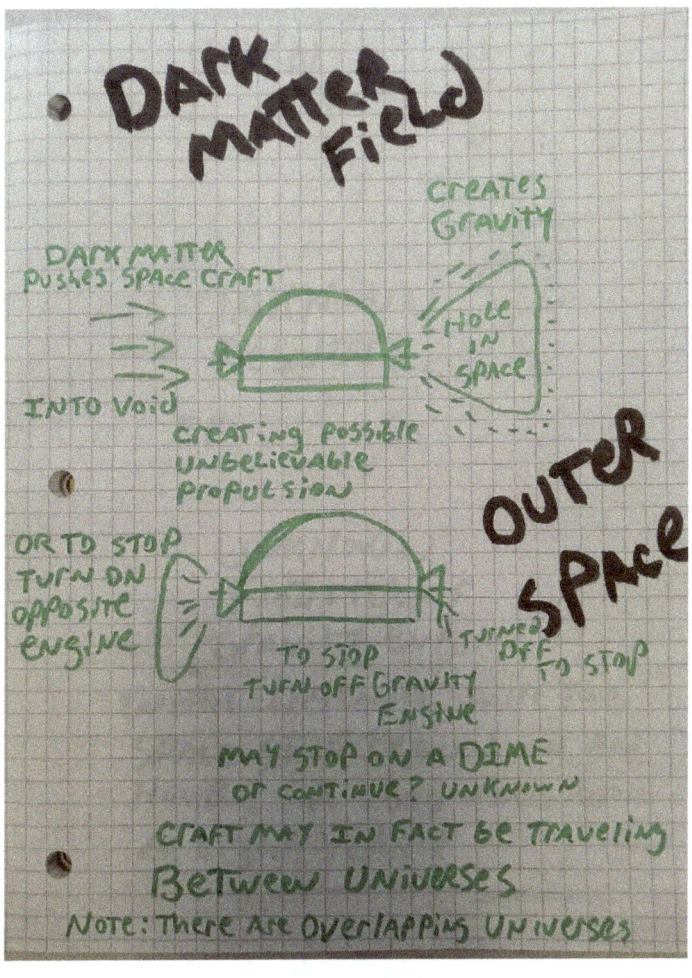

Unknown what type of propulsion? Yes, propulsion, yet possible enormous distances may be accomplished.

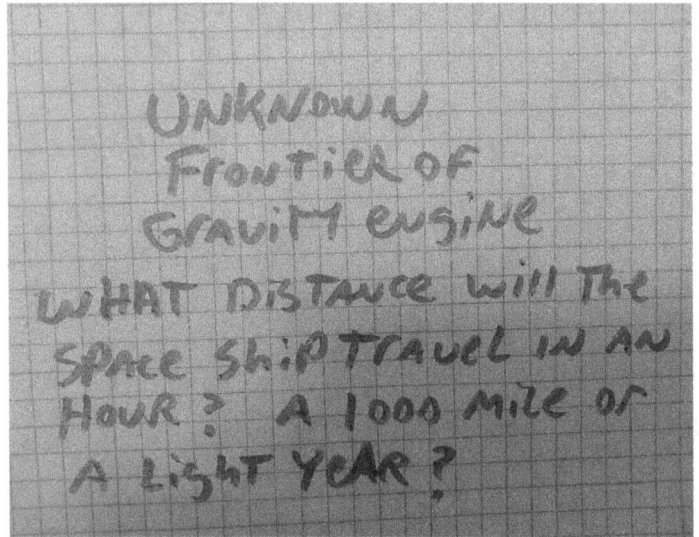

It is not normal space travel. It is being pushed into a Dark Matter field void. Does that mean travelling between universes as in a wormhole type travel? A wormhole not in the sense of a tube, but a continuing creating void. This can only be found out by testing. ***The idea of travelling a light year in a short time of a minute, or an hour, or a year, cannot be dismissed.*** Going to Mars or the Moon can be tested by aiming the Gravity engine and turning the engine on. In conclusion, the Gravity engine should be turned on **only for seconds at a time until the magnitude of this technology is understood.**

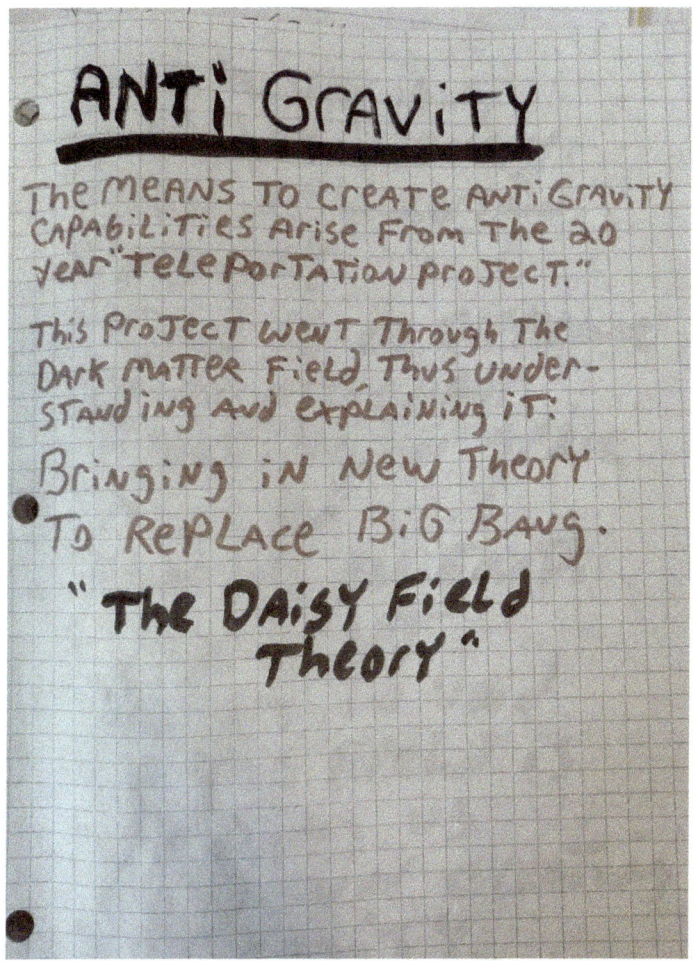

# ANTI GRAVITY

The means to create anti gravity capabilities arise from the 20 year "teleportation project."

This project went through the dark matter field, thus understanding and explaining it:

Bringing in new theory to replace Big Bang.

"The Daisy Field Theory"

Dark matter is a 3D vibration energy field at a certain frequency.

There is no difference of an astronaut floating in space

or

floating on Earth

except

we are in void + the field is pushing us down creating gravity.

## *What is the difference of this:*

## *and this:*

*As you can see a counter vibration wave is applied under the grass. The frequency will be exactly equal to the Dark Matter 3d vibration field. Thus neutralizing the push on any object regardless of weight. It could be a car, person, large stone unattached to ground: A new flying vehicles application.*

*Lift up an empty plastic bottle and drop it. That is the push down of the vibration field. That is what gravity is. Just as described with the ball in water. Now apply a machine under, or inside the bottle to emit an equal vibration to cancel out the field. Then the bottle will float.*

*I will mention, to put aside for now all of the effort in trying to find solutions to gravity using Poles of positive or negative. Yes, those are nice for applications of a few inches off the ground.*

*Totally remove those theories of magnetic applications and focus on just countering the dark matter field using vibration waves and a possible combo of electricity. Cannot someone in outer space just get a jar of empty dark matter field and close the lid. Then get a microscope to see what it is? Or at least measure its frequency so there is some starting point. It contains nothing? Possibly true, most likely being a vibration? May not even stay in the jar. Measure the vibration and you will have your answer to start a new era for humanity. The frequency to apply a wave against it is needed.*

*To try and use an example that sparked this breakthrough of creating anti gravity. Imagine the following apotheosis.*
*1. Outer space Dark Matter field can be imagined as particles standing still with an astronaut floating in the sea of the fundamental particles of the vibration field.*
*2. In Earth's atmosphere, the Dark Matter field can be imagined as a wave pushing us down in this same sea.*

# TIBETAN MONKS ANTI GRAVITY METHOD

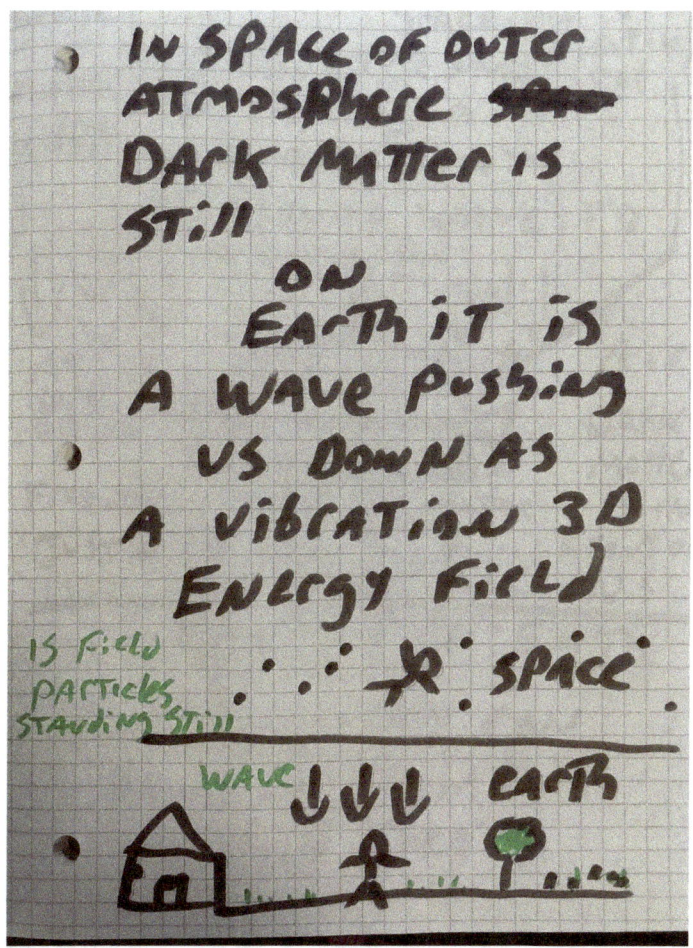

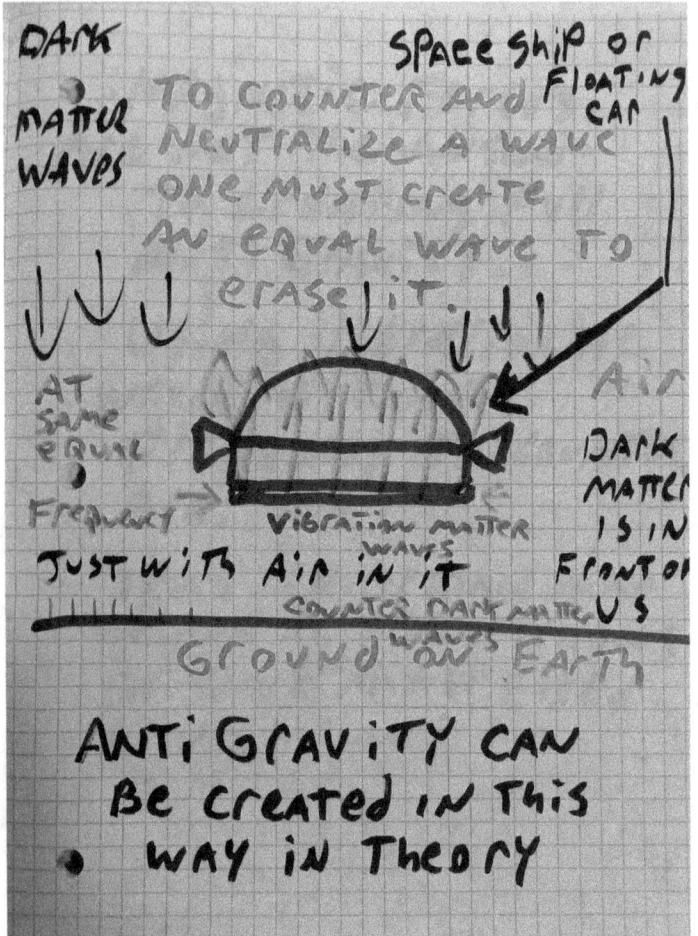

- Magnetic forces are not the answer to creating full anti gravity.

Using vibrational waves at exact frequency to counter are.

Tests in outer space may be able to find frequency of dark matter field of vibration?

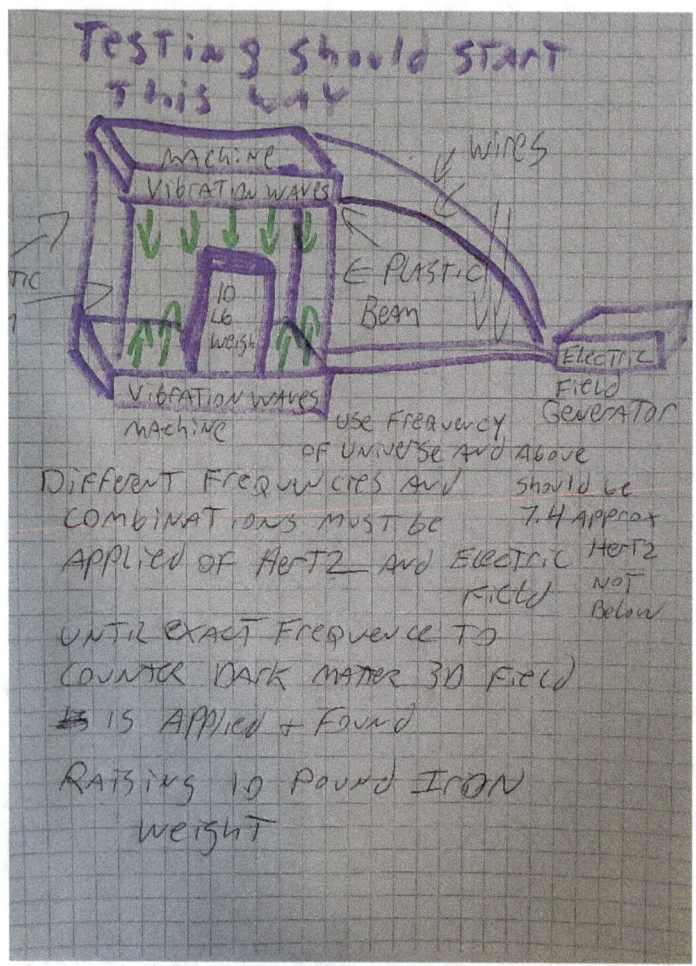

The visual of the test equipment of the 10 pound weight, portrays *vibration waves going up as a wave to counter the dark matter field wave going down.* However, since this is unknown, experiments can be made **using downward vibration waves as well as upward. Documenting all experiments in exact.**

To elucidate in the comparison (For example, **to create some kind of comparison)** of water as being Dark Matter: How is anti gravity created in water itself? It is accomplished by using an opposite of the same exact composition of the water. Water being H2O and breaking down the water itself into Hydrogen2 and Oxygen. Both it seems by *being an opposite equal force, but PUT BACK in the Water will repel the water: Canceling it. OR A BETTER TERM WOULD BE CREATING A VOID TO FLOAT.*

This new technology will be simular using the composition of the Dark Matter Vibration FIeld *being an opposite EQUAL force, but PUT BACK in the air or atmosphere will repel the Dark Matter Field: Cancelling it. OR A BETTER TERM WOULD BE CREATING A VOID TO FLOAT.*

**AUTOMOTIVE COMPANIES WOULD GREATLY BENEFIT FROM THE ANTI GRAVITY PAPERS.** The creation of floating cars would be a GLOBAL success for any advancing technology market.

**I would imagine since we do not feel any effect from the Dark Matter Field pushing us down, except gravity, The force wave to counter it would be just as suttle. Only having to be under the vehicle. It would not matter if an object weighed 100,000 pounds or 10 pounds. It is all just physics.**

# *TIBETAN MONKS LIFT STONES WITH ANTI GRAVITY SOUND FREQUENCY.*

"Demonstrates to a scientist of physics that a vibrating and condensed sound field can nullify the power of gravitation."

Tibetan Monks lift stones onto mountain top construction site, using sound vibration. These are 1939 sketches from Swedish aircraft designer Henry Kjellson.

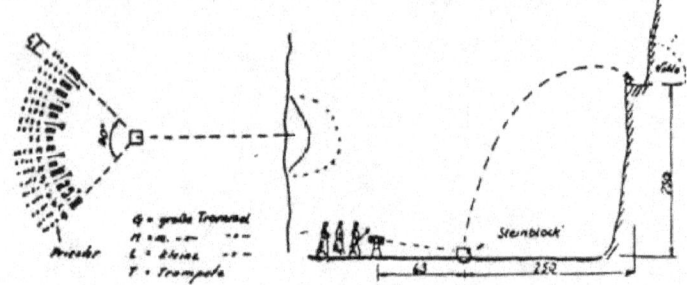

On the right of drawing is mountain side that the monks are buiding a wall. On the ground is a stone we will call "A" with a curved inside to house a large stone we will call "B" to be lifted into the air by soundwaves. The curved inside of stone "A" assists in the initial lift from the canceling out of the energy dark matter vibration energy field.

On the left are the monks and instrument musicians. T means >trumpeters S means>big drummers, M means>medium drummers, Inset shows method of suspending drum, and gives an idea of its size. Henry Kjellson writes in his journal:

**"The 200 priests are waiting to take up their positions in straight lines of 8 or 10 behind the instruments, 'like spokes in a wheel.'**

*"SWEDISH TRANSLATION TO ENGLISH of lower drawing: "Approx 200 priests are standing in rows behind the instruments."*

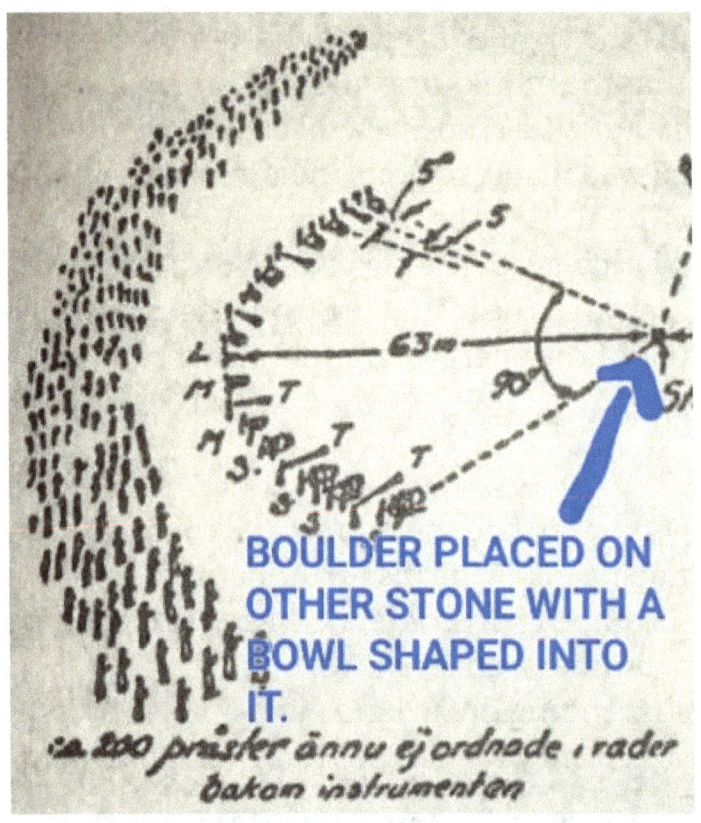

"We know from the priests of the far east that they were able to lift heavy boulders up high mountains with the help of groups of various sounds... the knowledge of the various vibrations in the audio range demonstrates to a scientist of physics that a vibrating and condensed sound field can nullify the power of gravitation."

Henry Kjellson was there as well as swedish engineer Olaf Alexanderson to take detailed measurements.

The following journal is from observing the monks raise the boulders. One after another. The German article "Implosion No 13" 1939 has Swedish engineer Olaf Alexanderson measurements as Henry made drawings and observations of this amazing scientific engineering human achievment.

This is his journal:
"A Swedish doctor, Dr Jarl, a friend of Kjelsons, studied at Oxford. During those times he became friends with a young Tibetan student. A couple of years later, it was 1939, Dr Jarl made a journey to Egypt for the English Scientific Society. There he was seen by a messenger of his Tibetan friend, and urgently requested to come to Tibet to treat a high Lama."

"After Dr Jarl got the leave he followed the messenger and arrived after a long journey by plane and Yak caravans, at the monastery, where the old Lama and his friend who was now holding a high position were now living."

"Dr Jarl stayed there for some time, and because of his friendship with the Tibetans he learned a lot of things that other foreigners had no chance to hear about, or observe."

"One day his friend took him to a place in the neighborhood of the monastery and showed him a sloping meadow which was surrounded in the north west by high cliffs. In one of the rock walls, at a height of about 250 meters was a big hole which looked like the entrance to a cave. In front of this hole there was a platform on which the monks were building a rock wall. The only access to this platform was from the top of the cliff and the monks lowered themselves down with the help of ropes."

"In the middle of the meadow, about 250 meters from the cliff, was a polished slab of rock with a bowl like cavity in the centre. The bowl had a diameter of one meter and a depth of 15 centimeters. A block of stone was maneuvered into this cavity by Yak oxen. **The block was one meter wide and one and one-half meters long.** Then 19 musical instruments were set in an arc of 90 degrees at a distance of 63 meters from the stone slab. The radius of 63 meters was measured out accurately. The musical instruments consisted of 13 drums and six trumpets (Ragons.)

Eight drums had a cross-section of one meter, and a length of one and one-half meters. Four drums were medium size with a cross-section of 0.7 meter and a length of one meter.

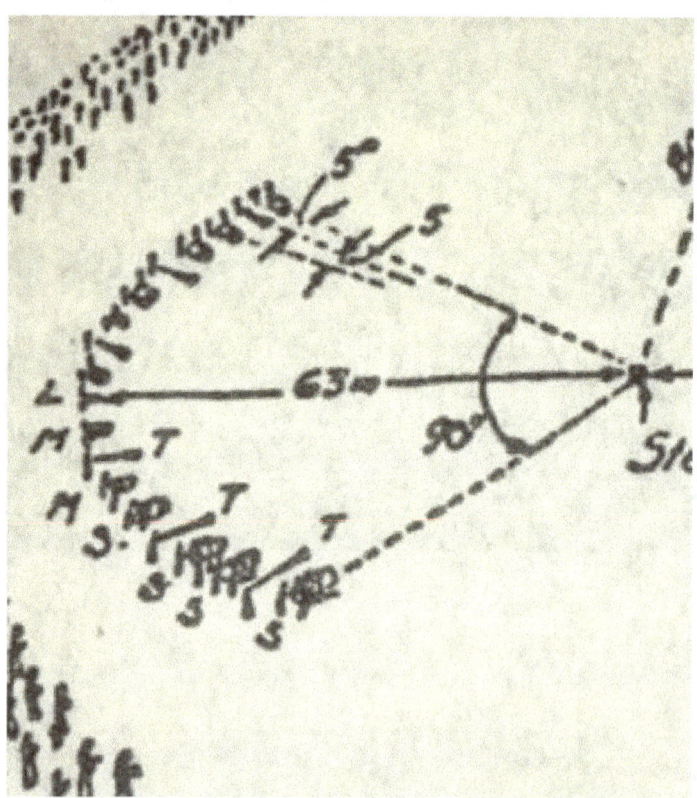

The only small drum had a cross-section of 0.2 meters and a length of 0.3 meters. All the trumpets were the same size. They had a length of 3.12 meters and an opening of 0.3 meters. The big drums and all the trumpets were fixed on mounts which could be adjusted with staffs in the direction of the slab of stone."

"The big drums were made of 3mm thick sheet iron, and had a weight of 150 kg. They were built in five sections."

"All the drums were open at one end, while the other end had a bottom of metal, on which the monks beat with big leather clubs. Behind each instrument was a row of monks. The situation is demonstrated in the following diagram:"

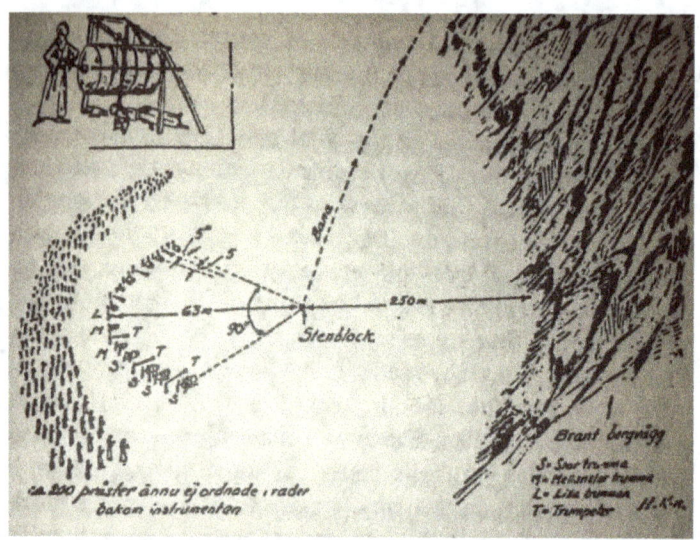

"When the stone was in position the monk behind the small drum gave a signal to start the concert. The small drum had a very sharp sound, and could be heard even with the other instruments making a terrible din. All the monks were singing and chanting a prayer, slowly increasing the tempo of this unbelievable noise. **During the first four minutes nothing happened, then as the speed of the drumming, and the noise, increased, the big stone block started to rock and sway, and suddenly it took off into the air with an increasing speed in the direction of the platform in front of the cave hole 250 meters high.**

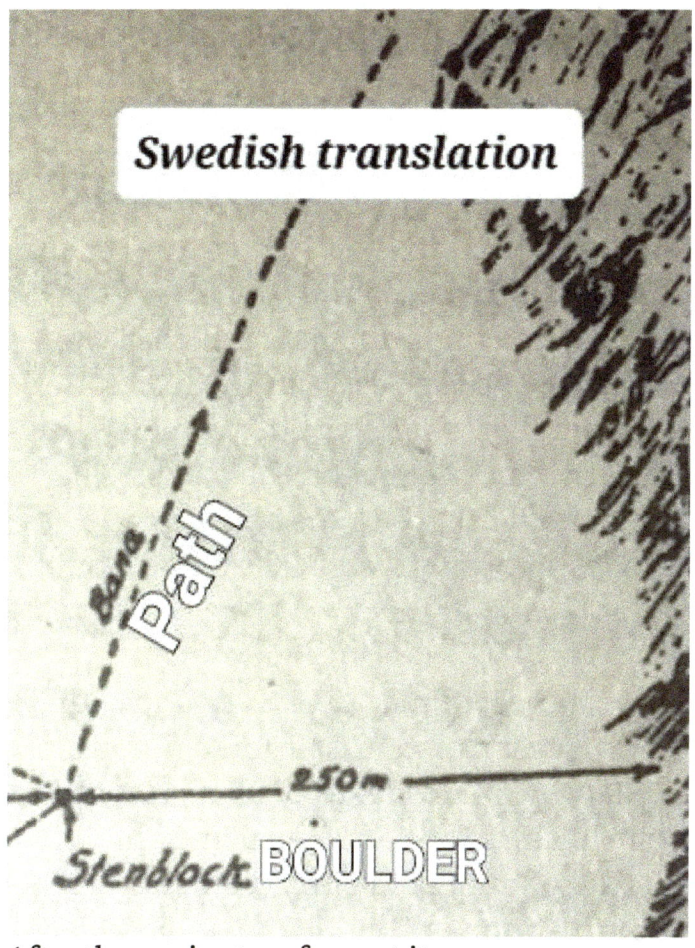

After three minutes of ascent it landed on the platform."

"Continuously they brought new blocks to the meadow, and the monks using this method, transported 5 to 6 blocks per hour on a parabolic flight track approximately 500 meters long and

250 meters high. From time to time a stone split, and the monks moved the split stones away. Quite an unbelievable task."

The article in 1939 concludes with stating the priests used their voices at different tones and increased with pitch at a steady chant. They believed the contents of the chant were of no gain, however the layout of the arc of the priest chanting at a sound wave combined with the instruments had the anti gravity effect on the stones. The cut out bowl shape in the bottom rock was also an inportant factor. The stone would raise the higher the pitch.

This has been filmed by Dr. Jarl and could be viewed if found. Last known film from 1939 was in 1990. The conclusion is these principles can be recreated using amplifiers instead of hundreds of instruments and chanting monks.

The ranges of sound hertz that can be started with for anti gravity tests, looking at the monks success combined with the new theory to explain and understand what gravity really is: Being an energy vibration wave, would be in the range of testing on a 10 pound weight of metal or stone: 5.1- 5.2- 5.3 -5.4 -5.45 5.49-5.5 Hertz in an arc or acoustic catching curved bowl with sound amplifiers around such a platform. Average being 5.4 hertz but these are unknown and must be tested and the frequencies of the instruments being note "C" 270 hertz and note "F" 5.4 hertz etc.. along the entire voice and intrument sound vibration sound spectrum, until the frequency vibration wave is found to cancel out the dark matter energy vibration field raising the stone or 10 pound iron weight for your experiment.

## THE THREE CROSSES IN A ROW UPRIGHT WILL BE OUR SYMBOL.

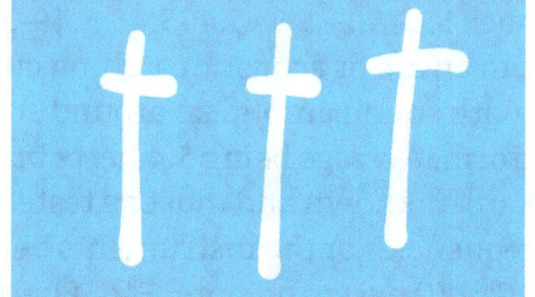

**That will be one of the symbols on these new gravity engine spaceships and floating cars.**

## WEAR THEM ON YOUR CLOTHES OR SKIN WHEN YOU TELEPORT TO OTHER WORLDS.

## WEAR THEM BECAUSE YOU ARE A PART OF A NEW CLUB TO ENLIGHTEN THE WORLD.

## WHEN YOU SEE ANOTHER WITH THE SYMBOL YOU WILL SMILE.

**MANY FRIENDSHIPS, MARRIAGES AND GROUPS WILL BE FORMED IN THIS WAY.**

**What better symbol for beings from other worlds to see. One cross isn't good enough for this exploration of the universe. Three are needed for full protection and light to others in the cosmos. The Father, The Son and the Holy Ghost.**

*When I was in "Another Land" I seen a girl who didn't have any hair. I gave her hair by wishing and focusing on it and verbally speaking matter into existence with my right hand out. And when the hair materialized on her head: It had three small silver crosses hanging from it.*

JOHNNY VINCENTO

# *ARTICLES OF THE MONTH FEATURING:*

Did ancients have fantastic technology to open dimensions with staffs that blast vibration waves?

*Theory is, that these staffs opened dimensions and that is what dry ground they walked on after water was separated. The key to opening dark matter field to worlds within worlds is vibrational waves.*

## *THE GOLDEN LOTUS STAFF*
### *EGYPT*

*Seneferu, father of the Pharaoh Khufu, reigned long over a contented and peaceful Egypt. He had no foreign wars and few troubles at home and with so little business of state he often found time hanging heavy on his hands.*

*One day he wandered wearily through his palace at Memphis, seeking for pleasures and finding none that would lighten his heart.*

*Then he bethought him of his Chief Magician,*

*Zazamankh and he said "If any man is able to entertain me and show me new marvels, surely it is the wise scribe of the rolls. Bring Zazamankh before me."*

*Pharaoh: "devise something that will fill my heart with pleasure"*

*Straightway his servants went to the House of Wisdom and brought Zazamankh to the presence of Pharaoh. And Seneferu said to him, "I have sought throughout all my palace for some delight, and found none. Now of your wisdom devise something that will fill my heart with pleasure." Then said Zazamankh to him, "O Pharaoh life, health, strength be to you! - my counsel is that you go sailing upon the Nile, and upon the lake below Memphis. This will be no common voyage, if you will follow my advice in all things."*

*"Believing that you will show me marvels, I will order out the Royal Boat," said Seneferu. "Yet I am weary of sailing upon the Nile and upon the lake."*

*"This will be no common voyage," Zazamankh*

*assured him. "For your rowers will be different from any you have seen at the oars before. They must be fair maidens from the Royal House of the King's Women: and as you watch them rowing, and see the birds upon the lake, the sweet fields and the green grass upon the banks, your heart will grow glad."*

*"Indeed, this will be something new," agreed Pharaoh, showing some interest at last. "Therefore I give you charge of this expedition. Speak with my power, and command all that is necessary."*

*Then said Zazamankh to the officers and attendants of Pharaoh Seneferu, "Bring me twenty oars of ebony inlaid with gold, with blades of light wood inlaid with electrum. And choose for rowers the twenty fairest maidens in Pharaoh's household: twenty virgins slim and lovely, fair in their limbs, beautiful, and with flowing hair. And bring me twenty nets of golden thread, and give these nets to the fair maidens to be garments for them. And let them wear ornaments of gold and electrum and malachite."*

*All was done according to the words of Zazamankh, and presently Pharaoh was seated in the Royal Boat while the maidens rowed him up and down the stream and upon the shining waters of the lake. And the heart of Seneferu was glad at the sight of the beautiful rowers at their unaccustomed task, and he seemed to be on a voyage in the golden days that were to be when Osiris returns to rule the earth.*

*But presently a mischance befell that happy party upon the lake. In the raised stern of the Royal Boat two of the maidens were steering with great oars fastened to posts. Suddenly the handle of one of the oars brushed against the head of the girl who was using it and swept the golden lotus she wore on the fillet, that held back her hair into the water, where it sank out of sight.*

*With a little cry she leant over and gazed after it. And as she ceased from her song, so did all the rowers on that side who were taking their time from her.*

*"Why have you ceased to row?" asked*

*Pharaoh.*

*And they replied, "Our little steerer has stopped, and leads us no longer."*

*"And why have you ceased to steer and lead the rowers with your song?" asked Seneferu.*

*"Forgive me, Pharaoh, life, health, strength be to you!" she sobbed. "But the oar struck my hair and brushed from it the beautiful golden lotus set with malachite, which your majesty gave to me, and it has fallen into the water and is lost forever."*

*"Row on as before, and I will give you another," said Seneferu.*

*But the girl continued to weep, saying, "I*

*want my golden lotus back, and no other!"*

*Then said Pharaoh, "There is only one who can find the golden lotus that has sunk to the bottom of the lake. Bring to me Zazamankh my magician, he who thought of this voyage. Bring him here on to the Royal Boat before me."*

*So Zazamankh was brought to where Seneferu sat in his silken pavilion on the Royal Boat. And as he knelt, Pharaoh said to him: "Zazamankh, my friend and brother, I have done as you advised. My royal heart is refreshed and my eyes are delighted at the sight of these lovely rowers bending to their task. As we pass up and down on the waters of the lake, and they sing to me, while on the shore I see the trees and the flowers and the birds, I seem to be sailing into the golden days either those of old when Re ruled on earth, or those to come when the good god Osiris shall return from the Duat. But now a golden lotus has fallen from the hair of one of these maidens fallen to the bottom of the lake. And she has ceased to sing and the rowers on her side cannot keep time with their oars. And she is not to be comforted with promises of other gifts, but weeps for her golden lotus. Zazamankh, I wish to give back the golden lotus to the little one here, and see the joy return to her eyes."*

*"Pharaoh, my lord, life, health, strength be to you!" answered Zazamankh the magician, "I will do what you ask, for to one with my knowledge it is not a great thing. Yet maybe it is a splendor you have never seen, and it will fill you with wonder, even as I promised, and make your heart rejoice yet further in new things."*

*Then Zazamankh stood at the stern of the Royal Boat and began to speak words of power.* ***And at that moment he held out his wand over the water, and the lake parted as if a piece had been cut out of it with a great sword. The lake here was twenty feet deep, and the piece of water that the magician moved rose up and set itself upon the surface of the lake so that there was a cliff of water on that side forty feet high.***

*Now the Royal Boat slid gently down into the*

*great cleft in the lake until it rested on the bottom. On the side towards the forty foot cliff of water there was a great open space where the bottom of the lake lay uncovered, as **firm and dry as the land itself.***

*And there, just below the stern of the Royal Boat, lay the golden lotus.*

*With a cry of joy the maiden who had lost it sprang over the side on to the **firm ground,** picked it up and set it once more in her hair. Then she climbed swiftly back into the Royal Boat and took the steering oar into her hands once more.*

*Zazamankh slowly lowered his rod,*

*and the Royal Boat slid up the side of the water until it was level with the surface once more. Then at another word of power, and as if drawn by the magician's rod,*

 *the great piece of water slid back into place,*

*and the evening breeze rippled the still surface of the lake as if nothing out of the ordinary had happened. But the heart of Pharaoh Seneferu rejoiced and was filled with wonder, and he cried: "Zazamankh, my brother, you are the greatest and wisest of magicians! You have shown me wonders and delights this day, and your reward shall be all that you desire, and a place next to my own in Egypt." Then the Royal Boat sailed gently on over the lake in the glow of the evening, while the twenty lovely maidens in their garments of golden net, and the jeweled lotus flowers in their hair dipped their ebony and silver oars in the shimmering waters and sang sweetly a love song of old Egypt.*

## *THE MOSES STAFF*
### *EGYPT*

**By Artist Andrea Andreani Italy 1558–1629**

*14 And the Lord spake unto Moses, saying,*

*2 Speak unto the children of Israel, that they turn and encamp before Pihahiroth, between Migdol and the sea, over against Baalzephon: before it shall ye encamp by the sea.*

*3 For Pharaoh will say of the children of Israel, They are entangled in the land, the wilderness hath shut them in.*

*4 And I will harden Pharaoh's heart, that he shall follow after them; and I will be honoured upon Pharaoh, and upon all his host; that the Egyptians may know that I am the Lord. And they did so.*

*5 And it was told the king of Egypt that the people fled: and the heart of Pharaoh and of his servants was turned against the people, and they said, Why have we done this, that we have let Israel go from serving us?*

*6 And he made ready his chariot, and took his people with him:*

*7 And he took six hundred chosen chariots, and all the chariots of Egypt, and captains over every one of them.*

*8 And the Lord hardened the heart of Pharaoh king of Egypt, and he pursued after the children of Israel: and the children of Israel went out with an high hand.*

*9 But the Egyptians pursued after them, all the horses and chariots of Pharaoh, and his horsemen, and his army, and overtook them encamping by the sea, beside Pihahiroth, before Baalzephon.*

*10 And when Pharaoh drew nigh, the children*

*of Israel lifted up their eyes, and, behold, the Egyptians marched after them; and they were sore afraid: and the children of Israel cried out unto the Lord.*

*11 And they said unto Moses, Because there were no graves in Egypt, hast thou taken us away to die in the wilderness? wherefore hast thou dealt thus with us, to carry us forth out of Egypt?*

*12 Is not this the word that we did tell thee in Egypt, saying, Let us alone, that we may serve the Egyptians? For it had been better for us to serve the Egyptians, than that we should die in the wilderness.*

*13 And Moses said unto the people, Fear ye not, stand still, and see the salvation of the Lord, which he will shew to you to day: for the Egyptians whom ye have seen to day, ye shall see them again no more for ever.*

*14 The Lord shall fight for you, and ye shall hold your peace.*

*15 And the Lord said unto Moses, Wherefore*

*criest thou unto me? speak unto the children of Israel, that they go forward:*

16 **But lift thou up thy rod, and stretch out thine hand over the sea, and divide it: and the children of Israel shall go on dry ground through the midst of the sea**.

*17 And I, behold, I will harden the hearts of the Egyptians, and they shall follow them: and I will get me honour upon Pharaoh, and upon all his host, upon his chariots, and upon his horsemen.*

*18 And the Egyptians shall know that I am the Lord, when I have gotten me honour upon Pharaoh, upon his chariots, and upon his horsemen.*

*19* **And the angel of God, which went before the camp of Israel**, *removed and went behind them;* **and the pillar of the cloud went** *from before their face, and stood behind them:*

*20 And it came between the camp of the Egyptians and the camp of Israel;* **and it was a cloud and darkness to them, but it gave light** *by night to these: so that the one came not near the other all the night.*

**21** *And Moses stretched out his hand over the sea; and the Lord caused the sea to go back by a strong east wind all that night, and made the sea dry land, and the waters were divided.*

**22** *And the children of Israel* **went into the midst of the sea upon the dry ground: and the waters were a wall unto them on their right hand, and on their left.**

**23** *And the Egyptians pursued, and went in after them to the midst of the sea, even all Pharaoh's horses, his chariots, and his horsemen.*

**24** *And it came to pass, that in the morning watch the* **Lord looked unto the host of the Egyptians through the pillar of fire and of the cloud,** *and troubled the host of the Egyptians,*

**25** *And took off their chariot wheels, that they drave them heavily: so that the Egyptians* said, Let us flee from the face of Israel; for the Lord fighteth for them against the Egyptians.

**26 And the Lord said unto Moses, Stretch out thine hand over the sea, that the waters may come again upon the Egyptians, upon their chariots, and upon their horsemen.**

**27 And Moses stretched forth his hand over the sea, and the sea returned to his strength when the morning appeared; and the Egyptians fled against it; and the Lord overthrew the Egyptians in the midst of the sea.**

28 And the waters returned, and covered the chariots, and the horsemen, and all the host of Pharaoh that came into the sea after them; there remained not so much as one of them.

**29 But the children of Israel walked upon dry land in the midst of the sea; and the waters were a wall unto them on their right hand, and on their left.**

30 Thus the Lord saved Israel that day out of the hand of the Egyptians; and Israel saw the

*Egyptians dead upon the sea shore.*

*31 And Israel saw that great work which the Lord did upon the Egyptians: and the people feared the Lord, and believed the Lord, and his servant Moses.*

## *THOTH'S STAFF*
### *ATLANTIS/PRE EGYPT*

*Notice the other piece in ground. Staff bottom*

*may connect to it. Thus activating staff.*

*Here we see that the **Staff** and this **Piece** work together as it is engraved to let us know. This original museum piece appears to*

*be made of copper. A depiction of a real detailed diagram of the unit. Not the usual general one of just an empty opening on top. They even gave us another image on object. The person with the Sun on its head. Which means " Positive Electrical Arc" or "Power of the Sun." The stacked slabs look as diods. "Regenerating Stability" the Egyptians called them. That is findings from looking into ancient egyptian light bulbs. My theory is that the object here was connected to a huge jar of a hundred or a thousand gallon baghdad battery jug full of vinegar. Connecting the staff to it would provide an unbelievable amount of power. How it was focused to vibration waves is unknown. Possibly Mercury was included in a hollow staff rod. Or it was attached to an obelisk. When the obelisk was hit with a giant hammer, the vibration of the obelisk would blast out of the staff. However, these don't match any of the three stories. A look at the obelisks should search for any electrical connecting areas. There could have been smaller portable devices to strike that would vibrate at extreme levels? Unknown. There are many ancient texts* **describing SOUND FREQUENCY from groups of monks voices raising stones.**

*Over the world then broke the great waters,
drowning and sinking,
changing Earth's balance
until only the Temple of Light was left
standing on the great mountain on UNDAL
still rising out of the water;
some there were who were living,
saved from the rush of the fountains.*

*Called to me then the Master, saying:
Gather ye together my people.
Take them by the arts ye have learned of far*

*across the waters, until ye reach the land of
the hairy barbarians,
dwelling in caves of the desert.
Follow there the plan that ye know of.*

*Gathered I then my people and
entered the great ship of the Master.
Upward we rose into the morning.
Dark beneath us lay the Temple.
Suddenly over it rose the waters.
Vanished from Earth,
until the time appointed,
was the great Temple.*

*Fast we fled toward the sun of the morning,
until beneath us lay the land of the children
of KHEM.
Raging, they came with cudgels and spears,
lifted in anger seeking to slay and utterly
destroy the Sons of Atlantis.*

***Then raised I my staff and directed a ray of
vibration,***

# TIBETAN MONKS ANTI GRAVITY METHOD

*striking them still in their tracks as fragments
of stone of the mountain.*

*Then spoke I to them in words calm and peaceful,
telling them of the might of Atlantis,
saying we were children of the Sun and its messengers.
Cowed I them by my display of magic-science,
until at my feet they groveled,* **when I released them.**

**And so it is to be, that with the correct matching frequency the following will be available to humanity using that same degree of vibration waves: The creation of anti gravity on Earth and the creation of gravity in outerspace.**

**Science update: If interested in refound teleportation. Image is ancient Star Map depicting teleportation from Earth to other planets in universe. The two planets**

are red in original. They are Mars. Yes two Mars. Two universes Two Orion's Belts. One Mars possibly overlapping another. Just like here on Earth. So it may be possible with the working Einstein Rosen bridge from "The Teleportation Project" to actually Teleport to Mars from Earth right now. An experiment should be made. I am too old to teleport anymore. It is best for people ages approx. 28 to 38. Teleportation to an alien planet or their spaceship may be attempted right now as well. Not fiction. For real. Read "Einstein Rosen Bridge now proven: Wormhole allows entire person to travel."

*Engineers:* *Look into the effects of vibration on Mercury. It may have be used for forever batteries, It is used for the forever lights, may be responsible for creating electricity and flying vehicles in Sanskrits and Ancient Egypt. Sanskrits even mention heating a Mercury engine to create whirlwind to raise vehicle off ground. The fumes are used as lights that stay lit for thousands of years (simular to fluorescent bulbs). The baghdad gigantic batteries may have used vinegar. However, Mercury in the lights batteries that stayed lit for thousands of years. Go to Amazon type Teleport News for the detailed evidence.*

*There is a truth that the West does not know about Jesus. He sent his disciple Thomas to India and his eight chruches are still standing today from around 2000 years ago.*

*Some Presidents of India even praise him in their speeches. The government even made postage stamps to commemorate him. Jesus himself studied Buddhism. This is a shocking truth, yet if you compare Buddha's parable of the seeds with Jesus' parable of the seeds 2500 years after Buddha, they are almost word by word.*

# ARTICLES OF THE MONTH FEATURING:

## THE RETURN OF MAITREYA - STOPPING THE WHEEL OF SAMSARA - THE END OF THE WORLD AND THE BEGINNING OF THE SATYA YUGA 2030AD.

Advanced study with reference numbers

"At that period, brethren, there will arise in the world an Exalted One named Maitreya, Fully Awakened, abounding in wisdom and goodness, happy, with knowledge of the worlds, unsurpassed as a guide to mortals willing to be led, a teacher for gods and men, an Exalted One, a Buddha, even as I am now. He, by himself, will thoroughly know and see, as it were face to face, this universe, with Its worlds of the spirits, Its Brahmas and Its Maras, and Its world of recluses and Brahmins, of princes and peoples, even as I now, by myself, thoroughly know and see them."
"Digha Nikaya, 26"
*reference #174*

## Saty सत्य

According to Buddha, the Dharma would disappear 2,500 years after His Dispensation began (1957 A.D.), and shortly after, the Maitreya would come.
"Abhidharmakosha, 4.12c. III. p. 41" "The monks and stream attainers will be strong in their union with Dharma for 500 years after the Blessed one's Parinirvana. In the 2nd 500 years they will be strong in Meditation; in the 3rd 500 years they will be strong in erudition (great knowledge.) In the 4th 500 years period they will only be

occupied with gift giving. The final 5th cycle of 500 years will see only fighting and reproving among the monks and followers. The pure Dharma will then become invisible - very little spirituality"

The Maitreya will arrive after the Kali age (age of darkness), which would be sometime after August 1st, 1943. Around 2,500 years after the beginning of the Dispensation (LIFE TIME) of the Buddha, the Maitreya will arrive. This would be around 1957 A.D.

The year of 1957 is based on adding 2,500 years from 544 B.C., one possible date for Buddha's death / beginning of Buddha's Dispensation/ the "Buddhist era." Another date that is possible for Buddha's death is 483 B.C. to begin this *reference #174* countdown. So adding 2,500 years to 483 B.C. gives the year 2017. Maitreya is expected to arrive at this time in history - around the year of 2017. Which means he is in the world now.

"Long before this day is going to pass and in that time of the people with an eighty-thousand-year life-span, there will arise in the world a Blessed One, an Arahant fully enlightened Buddha named Maitreya, endowed with wisdom and conduct, a Well-farer, Knower of the worlds, incomparable Trainer of men to be tamed, Teacher of gods and humans, enlightened and blessed, just as I am now."

## J.V. जॉनी विं सेंटो

Saty, Is this 80,000 a mistake??
I got it!!! The "80,000 year lifespan" is symbolic for top enlightenment. The other sayings say a "10 year life span" is symbolic for zero enlightenment. No one has either of those life spans, but those numbers have meaning symbolically.

It seems Buddha was referring to the mind of a person in Heaven with the knowledge of a person living 80,000 years (that lady I met in an overlapping world said she was 50,000 years old) and on earth many don't have spiritual knowledge more than a 10 year old, (even if they are in their 70's) 80,000 years means the knowledge of what *reference* **#175** existence is, Heaven and enlightenment. 10 years means a person has only the knowledge of being alive and their physical environment. They were born and believe in nothing and when they die they disappear. As was they have no memory of before being born, that is their belief for after they die: Nothingness. So this is why people do evil, Because they think there is no consequences of Hell. There is a guy who went to Hell. Jesus showed him. He told his story on tv. So he can tell the world that that place exists. The book is called "23 minutes in Hell."

I worked with these grown males in their 20's 30's 40's 50's 60's 70's and even talked to men in their 80's and most are not men, just in grown bodies with a grade school mind and emotions.

The reason I was arrested so many times was

because of fighting. They are demons who look for violence and I was foolish enough to give it to them. Didn't Buddha say don't ever fight if you gain nothing. Someone said those words? It is lonely to be among children and even when I won I lost. I longed to be around grown humans who have evolved as in the East. That is why I seeked to have students and friends from India. I don't belong in the West.

*reference #176*

### Saty सत्य

Guru, The thing is, it was said that when Jupiter, Earth and Sun will come in one line then there will be beginning of Satya Yuga. Satya Yuga is known as the age of truth, when humanity is governed by gods, and every manifestation or work is close to the purest ideal and humanity will allow intrinsic goodness to rule supreme. It is sometimes referred to as the "Golden Age".

The Satya Yuga, in Hinduism, is the first and best of the four yugas in a Yuga Cycle, preceded by Kali Yuga of the previous cycle and followed by Treta Yuga.

THE KALI AGE IS BEFORE THE SATYA AGE. This is what earth is getting out of, The Kali Yuga (Sanskrit: किलयुग, romanized: kaliyuga or kali-yuga) means "the age of Kali ", "the age of darkness", "the age of vice and misery", or "the age of quarrel and hypocrisy". Kali Yuga is described in the Mahabharata, Manusmriti, Surya Siddhanta,

Vishnu Smriti, and various Puranas.

### J.V. जॉनी विं सेंटो

So my friend looks like we are in or just got out of the age of vice and misery and no morals.
What is the duration of each of the four yugas or ages according to Buddhistss and Hindus?
*reference #177*

## Saty सत्य

The four great epochs in Hinduism are Satya Yuga or the Age of Truth is said to last for 4,000 divine years, Treta Yuga for 3,000, Dwapara Yuga for 2,000 and Kali Yuga will last for 1,000 divine years —a divine year equalling 432,000 earthly years.

Kali Yuga has two phases: In the first phase, humans—having lost the knowledge of the two higher selves—possessed knowledge of the "breath body" apart from the physical self.

Now during the second phase, however, even this knowledge has deserted humanity, leaving us only with the awareness of the gross physical body. This explains why humankind is now more preoccupied with the physical self than any other aspect of existence.

Due to our preoccupation with our physical bodies and our lower selves, and because of our emphasis on the pursuit of gross materialism, this age has been termed the Age of Darkness—an age when we have lost touch with our inner selves, an age of profound ignorance.

Another theory interprets these epochs of time to represent the degree of loss of righteousness in the world. This theory suggests that during Satya Yuga, only truth prevailed (Sanskrit Satya = truth). *reference #178* Sir Explanation of Yugas During the Treta Yuga, Earth lost one-fourth of the truth, 3/4 truth left in the world.

During the Dwapar Yuga Earth lost one-half of the truth, 1/2 truth left in the world and now the Kali Yuga is left with only one-fourth of the truth. Evil and dishonesty have therefore gradually replaced truth in the last three ages. Earth is left with 1/4 truth left in the world

## J.V. जॉनी विं सेंटो
What Happens Next?

**Saty** सत्य

According to Hindu cosmology, it is predicted that at the end of the Kali Yuga, Lord Shiva will destroy the universe and the physical body will undergo a great transformation. After the dissolution, Lord Brahma will recreate the universe, and humankind will become the Beings of Truth once again.

## J.V. जॉनी विं सेंटो

These vast years may be easily explained by the fact that time is different in other planes of existence. When I was in heaven I asked "How long will I live?" the answer was "217" or "two seventeen" to be exact. Because the words were

**179** spoken there could be a clue here. At that time I was 38 years old. When a Yuga is 430,000 years long for example and the Gods consider one year of human Earth which resides on a totally different plane of existence, Dividing 430,000 God years / 360 Human years gives 1200 Human years. It is just an idea to look into. for time is different on other planes. I asked a woman in one of the overlapping worlds. Which level is unknown, how old she was and my Father in Heaven was there. She answered "50,000 years." So this only means to take a new look at understanding these large years.

### Saty सत्य

The ages in ancient Indian scriptures is mentioned according to Vedic astrology. Western calendars are been used today so nobody uses Vedic calendars to keep track of these things. The thing is it was said that when Jupiter, Earth and Sun will come in one line then there will be beginning of Satya Yug and according to western calendars this happened in 1943 as you mentioned.

The mixture of Western and Vedic calendars have became a giant mess. Astrologically we can know what will be after 4 lakh 27 thousand years but western calendars cannot calculate that. That's what I am trying to say sir invaders have made a big mess of our ancient texts.

**180**

## J.V. जॉनी विं सेंटो

In Buddhism, Maitreya is the Fifth Buddha, in Hinduism - Kalki Avatar, the incarnation of Lord Vishnu. "The Secret Doctrine" states that "Maitreya is the secret name of the Fifth Buddha, and the Kalki Avatar of the Brahmins (Gods) - the last Messiah, who will come at the culmination of the Great Cycle." "The only difference lies in the dates of their appearances. Thus, while Vishnu is expected to appear on his white horse at the end of the present Kali Yuga age "for the final destruction of the wicked, the renovation of creation and the restoration of purity," Maitreya is expected earlier," – supplemented by H. P. Blavatsky in the "Theosophical glossary." "The Secret Doctrine" teaches that "the simple secret is this: there are cycles within greater cycles, which are all contained in the one Kalpa of 4,320,000 years. It is at the end of this cycle that the Kalki Avatâra is expected."

Maybe this last explanation will help with your dating system alignment.

## Saty सत्य

In this future epiphany or tenth avatar, the heavens will open and Vishnu will appear "seated on a milk-white steed, with a drawn sword blazing" like a comet, for the final destruction of **181** the wicked, the renovation of 'creation' and the 'restoration of purity'"... This will take place at

the end of the Kali Yuga 427,000 years hence.
Yes sir that's what I am trying to say. No matter
Buddhism has became a different religion the root
remains the same.

## J.V. जॉनी विं सेंटो

Could it be that the ancient India writings
could align with the arrival of KALI AVATAR the
punisher of the wicked and redeemer of the good
in the year of 2030 AD just 9 years from now?
(Time Travel for real)

I didn't want to upset you and didn't tell you that
I did see the future a second time and was taken
there half physical half vision, but was there able
to speak hear touch and even taste and feel some
water I was drinking. The guy sitting next to me
said "This may be the extinction of the human
race, we only have enough food and water for 3
days, we need to get to the Southern states." We
stood in the vision or teleportation in the future
and I said "What year is it?" He replied "2030" and I
said "That's right last year was 2029" teleportation
at its extreme for mankind. Yet not my doing in
the least but the "Spirit of Truth" who showed
me things to come. Might be just a nuclear war.
There is a mention of a huge 7 day war prophecy
I read from Buddha writing **182** that will precede
the coming of Maitreya. It has to be one or the
other. I seen a rural city abandoned so people
had time to flee. I was actually there and heard a
strange machine outside the building I was in that

sounded like "UMMMMMMUM, UMMMMMMUM, UMMMMMMUM." Maybe a drone?
This info was given to a Congressman in 2013 AD and he then passed the info to NASA.
When someone sees the future they are not in the future but the present. And now that you have taught me India prophecy it seems that KALI AVATAR is going to be arriving on a WHITE HORSE (asteroid comet or nuclear missile) with a BURNING SWORD into the den of EVIL DOERS in the NORTHERN top of the USA in 2030. For in America is the lowest of YUGAS. Visions from the "Spirit of Truth" don't lie. It is not going to happen in 2029 or 2031 or any other year only 2030. At least for this vision of time travel. Maybe that is what it takes for people to believe our words. Just the year I told you because I asked.

**Saty सत्य**

WOW! You are a true Guru, that is amazing and I am not worried because you mentioned Nostradamus said the East (SUN) will be around **183** for the next 7000 years to carry our message. Kalki avatar will not show himself after being born. The thing is after he is born he will gain knowledge from the immortal people like PARSHURAM (6th incarnation of Lord Vishnu), HANUMAN (One of the incarnation of Lord Shiva), VED VYASA (One of the incarnation of Lord Vishnu), ASHVATHAMA (One of the incarnation of Lord Shivs), etc. he will gain every knowledge and he will be the only person in Kali Yuga to have the 100% knowledge of the Vedas. But it would be an honor if he learned from our book too of this teleportation that you refound.

## J.V. जॉनी विं सेंटो

Someone has to bring back ancient teleportation to India. For me the honor would be if ISRO worked with us with starting Time Travel experiments as well as my DIMENSION SEPARATION VIEWER theory. The last place I would ever want to work is the USA. This country is one of the most unspiritual places on Earth. Saty, did you know that is normal for people in the USA to have animal heads hanging in their homes on their walls with their kids running around like it is normal? Not only that, they kill the animals just to have them stuffed. So the animals die so the evil doer can have it as a decoration. What a den of darkness this country is. My Dimension Separation Viewer could **184** change the world too by duplicating the window to Heaven that is described in the Bible "The road to Damascus." This would prove, in an actual physical tangible machine of engineering, to all the people in the world, that there is a Heaven and if you are of LIGHT of goodness and pure, you go to Heaven. A paradise. The LIGHT coming out of the window shining from Heaven will change Homo Sapiens into an evolved New species by actually advancing, opening, changing the brains of humans passing their dna to a new race of super enlightened beings, bringing in SATYA YUGA for real with a thousand years of peace or forever. The LIGHT changed and enlightened me instantly Saty. Imagine if I had more doses of that LIGHT?

**Saty सत्य**
I would like to live in that world and see that machine and the light.

### J.V. जॉनी विं सेंटो
My friend, Are you saying that Buddhism and Hinduism agree on the next Buddha coming while we live in our era? "Metteyya" and "Kali Avatar" are the same person reincarnated?

**185**

**Saty सत्य**
Yes. Every incarnation of Lord Vishnu is the same.

### J.V. जॉनी विं सेंटो
PLANET ALIGNMENTS TO DATE RETURN
I went on the site that aligns planets past and future. A vision of the future or past can be dated. That is why the ancient writers put down the alignment. However, numerous dates will show same alignment over time past or future. You can go back in time and actually date any alignment ancient writers are documenting.
Events such as visions are dated to the star

alignment of the day the vision was seen. This
could use more evidence, but I tested it. The date
of going back in time was March 26, 2006. I went
back to when I was a child of about 5 or 6 years old,
looking out of my own eyes for a half second. I was
holding a Native American toy totem pole that I
had as a kid. So I searched the planet alignments
for that day March 26, 2006. I went back week by
week day by day until I found the MATCH. All the
planets didn't line up, but I overlapped March 26,
2006 with the first month of **186** January 1976.
There was an exact overlay of the planets Mars,
Venus and Saturn. So that would put my age I
Time traveled to at 7 years old. No other dates
match even close. I looked from when I was 5 to 7.
I acquired the toy when I was 5 or 6 so that was the
years to search. Couldn't find a match so started
looking when I was 7 and that was Jan 1976.
PLANET ALIGNMENTS for birth of Kali Avatar,
Please conclude our last subject by finding
the planet alignments that were for birth of
Kali Avatar. With a date for our present
Calendar. If he is to be the same incarnation as
the 5th Buddha then the date is our time. Using
the computer program of planet alignments

**Saty सत्य**
Kalki avatar will be born when Moon will be in
Dhanishta Nakshatra, and the Sun will be in the
Swati Nakshatra.

### J.V. जॉनी विं सेंटो

This conversion to the western computer Star map program will be placing the MOON in the 23rd NAKSHATRA (DHANISHTA) being 23° 20' CAPRICORN to 06° 40' AQUARIUS...........Then place the SUN in the 15th NAKSHATRA (SWATI) being 6° 40' to 20° 00' LIBRA or (TULA) the SUN will be completely in LIBRA.

This will take some time for you, but it has to be done. We will always wonder what is the answer if it isn't calculated.

### Saty सत्य

Now according to these calculations if you will search on internet then you will get the date 2012. But I don't know much about it. I know one thing that he will "APPEAR" at the end of Kali Yuga. Guru, but I don't think that it will predict it accurately. So is the date 2012 the end of Kali Yuga?

### J.V. जॉनी विं सेंटो

In a way Saty, this book we are making for the world is helping bring in Satya Yuga. An exact date is not needed, 2012, 2017, 2030, it's all the same time frame concidering.

**Can ALL present and previous lives Karma be erased forever? Stopping the Wheel of Samsara?**

**What is the ancient secret of Jesus Christ to save India and all of Asia?**

*IT IS THAT JESUS BAPTISM ERASES ALL PREVIOUS LIVES SINS KARMA - STOPPING THE WHEEL OF SAMSARA.*

"Maitreya is the one recognized by all directions of Buddhism. A Bodhisattva which is the enlightened one who vows to

become a Buddha.
A man achieving the highest level of spiritual ability, he sees a world of endless reincarnation and evil."

"He is a messenger as to help other human beings come out of the "Wheel of Samsara," the endless circle of rebirths and sufferings."
*reference #189*

## Raaja राजा

Sir, I read the draft you sent me for your new book. WOW!!!!! You bringing a possible way to end karma and these never ceasing rebirths into this world is really something monumental. This concept is most lovely for India.

I would add though, there are 3 Karmas: 1st Sanchita Karma (karma of past lives) 2nd Prarbda karma (Karma the soul decides to clean in this life) 3rd Kriya karma (karma we do in this life)

## Saty सत्य

Guru, the problem is people don't understand and don't believe in these things so there is no research on these things for physicists to even study. People just believe what they see (the physical universe) but in fact what we call reality **190** is not really existing it's just an illusion in which we are trapped.

## J.V. जॉनी विं सेंटो

Here is a "Theory of all theories" and is a great parable for Karma as well. This theory can only be answered by asking questions to the immortals across dimensions.

Not an illusion my friend, unless you want to believe that the world is already ended and our spirits are put into past life physical recordings to live. That would go great with Karma. But our spirit thinks we are alive in our body. But it has already happened and is just a physical recording

that a spirit can be born into throughout all time.
Even our thoughts then would be recordings.
The bad person killing people in the first original
world (that was not a recording) now can have his
spirit put into the bodies of the people he killed.
The Police officer who falsely sent a man to life in
prison now can be born into the life of the one he
put in prison to live.
JUST A THEORY BUT COULD BE A POSSIBLITY. If
the Earth is destroyed, spirits will have to live in
past time. That would be an illusion.

### 191

### Saty सत्य

Yes that's right sir. I think that Time is a cycle
what we have done what we are doing and
what we are doing has happened. Our future is
determined from the beginning of the universe.
Today we are discussing, this discussion has
happened many times. Earth has been created
and destroyed many times. I think if a person
dies his soul enters another body according to
his karma and a person who is pure of heart and
Yogi gets free from the cycle of birth and death
and his soul travels and merges inside the most
supreme energy which we call as god. Our future
cannot be changed just like our past because it
has happened already. If we could change future
or past then it could create a paradox which could
make our universe a big mess. But it's difficult
to predict until we ourselves experience it.

## J.V. जॉनी विं सेंटो

You live in INDIA where even Buddha may have said Karma cannot be erased. This is not true, Buddha was wise, but Karma can be erased. It is true when Buddha was alive karma could not be erased. Another God came after him called Jesus. That is my message to India. You can do an experiment: Find someone who has a big problem with his payback Karma. Get that person Baptized (full SUBMERSION) taking Jesus as his **192** lord and savior (Buddha was 500 years before Jesus). Some stories between them are identical word to word almost. This doesn't mean getting rid of Buddha's Dharma. Before going under the water say "I REPENT FOR ALL SINS I DID IN PREVIOUS LIVES. I REPENT FOR ALL SINS I DID IN THIS LIFE. THE PAYBACK IS OVER, THE KARMA ENDS NOW, IT'S OVER WHEN I GO UNDER THE WATER!!!" and that is how you get rid of Karma. Test it on the person with the worst Karma. The person must believe and be truthful in a new life.

If one is happy with the results follow Jesus fully. When I teleported to Heaven on 03 26, 2006, I asked my God:

"Father how long will I live?" "Two seventeen" "I am much older than seventeen" "everyone has an inner age" "You mean like an emotional age" "yes" "but I want to know how long I will live?" "Don't worry you have many numbers"

On December 04, 2016 I was baptized. The old me

died and the new me is the one who tries to live a good life reborn in living my life as Jesus taught or as best I can. So it looks like the old me, "My inner age" wasn't going to live to the year 2017.

**193**

### Saty सत्य

And as you said Guru, about erasing karmas, if one's karma is erased then his motive of life is finished. We take birth and die due to karma but if main cause of life which is karma are erased then that person achieves Moksha that is free from birth and death which means he will never be born again

### J.V. जॉनी विं सेंटो

And If I did find a way to erase my Karma forever because the payback stopped when I got baptized then I found the way to not come back to earth anymore and that is fine with me. Because another thing I said to God and about a hundred people watching before I went under the water was: "Father in Heaven when I die please don't bring me back to this world, put me in level 2 or in Heaven if I am needed"

### Saty सत्य

Yes sir that's right but Karma cannot be erased. Sir you are right but what you are talking about is "Paap and Punya". Paap means sins or bad things that we do and Punya means good things that we do. Paap and Punya can be erased just as you said Baptism, in India River Ganga is the **194** most Holi river which is said to be arisen from Lord Shiva's Jata (Coiled hair), here it is said that if we bath or drink water of River Ganga all our Paap (sins or bad things) are washed out just similar to baptism. But sir karma is a very different thing what we do every day in fact what we do every micro second is counted in our karma. We are discussing about different things this is also karma. We sit, we walk, we eat, we sleep every single action is karma. Karma is the only thing which remains with us after death so we take birth every time. Lord Krishna said that doing and karma selflessly and without expecting anything from anyone, one can attain Moksha (free from cycle of birth and death).

In fact like lord Krishna, lord Buddha is also an Avatar (incarnation) of Lord Vishnu.

### J.V. जॉनी विं सेंटो

It erased my Karma. Let's try. Find a person who is getting very bad (payback karma) in this world. Just a person who knows that he or she is being punished. Then tell that person to do what I described. If that works do a group of ten for the experiment. Adding Jesus to this erasing karma won't take anything away from India's religion or an individual.

**195** You said last year that Hinduism is ever changing like new branches of a tree. But if it works one must then follow Jesus fully. That would be a new tree, I will show you that Jesus's Dharma is the same as Buddha's. Jesus just had power from the universe to erase karma. Since I teleported many times I would qualify as a SAGE with knowledge past this Earth. Basically, it is a test of which God is going to erase a person's karma. Not the baptism of your history/ a new baptism of saying before going under the water "I take Jesus as my lord and savior, I repent for all my sins in this world and in any other lives I did in the past, the karma is over, no more payback, the karma ends when I go under the water!"

You must at least see that I am a scientist that is unique in trying to help out the people of India. So do your part with one simple experiment. It's not for you or me. It is for billions of people in India's future.

Also, that person you are telling this too must not have any doubt that it will work. And YOU must not have any doubt. This saying it won't work and it can never be erased is bad thinking. I am a new scientist. If you believe my Teleportation then this theory of erasing Karma had success with me. So it has enough evidence to be tested. (Erase the past karma, this is new science) Even though it **196** sounds like religion (religion is science) even karma is science, for worlds are separated according to what degree of karma (LIGHT or shade of darkness one has in them: Negative or Positive Vibration.)

**Saty सत्य**
Yes sir I truly believe in your teleportation experiments. But this is very sad truth that karma cannot be erased. Yes we can erase our bad sins but not karma. And yes we can even reduce the effects of karma but cannot erase it completely. Erasing karma makes our soul completely 100% divine that means we are free from cycle of time and we go to the most supreme energy which is very very hard to achieve. Ones there are a saint in India and lord Vishnu sent his own vehicle the Garuda (Griffin) to take him to Vaikuntha (Heaven of lord Vishnu) this was because he did good karma. Doing good and selfless karmas is in fact reducing the effect of bad karma. No matter how divine we are, we have to suffer for our bad sins and yes only it's effects can be reduced through good and selfless karma but cannot be erased. Karma is such complicated and dangerous that even incarnations of gods were not able to escape from it. Today we know each other and this is also result of karma.

**197** Every person in our life, our family, our friends, our teachers are due to our as well as their karma. The root of every smallest cause in our life is due to karma. Today we are born on Earth because of our bad karmas in past life. This is so bad that we cannot erase our karma.

### J.V. जॉनी विं सेंटो

Very impressive knowledge, you remind me of Jesus's disciple Thomas, he was always doubting everything too. When Jesus told him to go to India to preach, Thomas said he didn't want to go. Can you believe that? So now let us join our research and knowledge to free India's people from karma. Yes I heard what you said and we will have to look at it from a different point of view.

### Saty सत्य

Sir, I know you are a great scientist, for I believe your work because India's past had Teleportation as well as time travel. You just rediscovered it. But, DO YOU KNOW WHAT IS KARMA?

### J.V. जॉनी विं सेंटो

It is action. And it is a recording of all your past lives all stacked on a person's spirit. An endless balance sheet of debits and credits, the CREDITS GOOD the DEBITS bad. A never ending cycle of rebirth to AT LEAST TRY to even the sheet. Then **198** when the sheet is evened a person has a chance in that life. To live as purely, physically, mentally, spiritually, to rise to PURE LIGHT on all of these. Then one achieved MOKSHA. And will not have to be reborn and can finally go home.
Few, very few reach MOKSHA.
Jesus came to give us the ritual to be REBORN to even the sheet of credit and debit in a present life and to live a life of LIGHT to achieve

MOKSHA. So the BAPTISM and REBORN is the same principle as Hindu search for MOKSHA.

**Saty सत्य**
SIR THROUGH THIS YOU ARE DIRECTLY QUESTIONING THE HINDU RITUALS.

**J.V. जॉनी विं सेंटो**
No, I would never do that, just adding one. Hindu's never added what Jesus brought. Do You agree that Jesus was a God?

**Saty सत्य**
I KNOW THAT JESUS WAS A GOD.

**J.V जॉनीवंसेंटो**
My friend, didn't that God have a message. What was the message? It agrees with seeking MOKSHA. HE SENT SAINT THOMAS HIS **199** DISCIPLE to India. The Churches are still in India today as we speak. There is eight of them.

Jesus talked about Heaven being a paradise that is right in front of our faces, but we can't see it. I was there on several occasions and it is wonderful.

The reader can decide if they want to try the experiment by adding Jesus ritual. If it works we will have freed billions of people in the future by allowing at least A CLEAN START to try and achieve MOKSHA in that person's present life. Whoever wants out of this endless rebirth and wants to try erasing their Karma just to start to achieve Moksha contact a Christian church in

India or the Saint Thomas Churches that have been there since the time of Jesus. Behind one of Saint Thomas's original churches is a new huge beautiful church.

The main thing is, as soon as you erase all your karma and come out of the water, karma starts all over again. That's when you have a chance to work on achieving Moksha.

**Saty सत्य**
AND CAN YOU TELL ME HOW TO GET TO THE POINT OF ACHIEVING MOKSHA?

### J.V. जॉनी विं सेंटो

Saty, here is the way, It is just an experiment and it worked for me.

Don't we bring both sides of the question to the reader? The reader can decide if they want to try the experiment by adding Jesus ritual.

**1.** The ritual of Jesus, is being REBORN in this present life while still alive and all their sins die with their old soul.

**2.** Jesus said many times "We are born in Sin" why would someone who doesn't know karma say such a statement? and he actually died for our sins. He died so we could erase our sins and go to the Kingdom of Paradise.

**3.** The ritual or correct Baptism for erasing all Karma, is to have person doing the baptism put olive oil on one's forehead in a cross or pour on one's head for the first seal of the HOLY GHOST: Then before going under the water say "I repent for my sins and take Jesus Christ as my Lord and Savior. I repent for all my sins in this life and all previous lives." The one doing the baptism will say "I baptize this person "AND SAYS THEIR NAME" in the name of the Father and the Son and the Holy Ghost." This is the first seal.

**201**

**4.** Once going under the water the old you, your old soul that is stacked with Karma DIES ALONG WITH ALL THAT KARMA in GOD'S NAME JESUS CHRIST.

**5.** Now this God Jesus gives you a clean start, a clean slate.

**6.** At this moment you are free of all Karma and YOU ARE REBORN WITH A NEW SPIRIT. By this I mean you are Baptized both in the name and authority of God Jesus and the authority of the Holy Ghost. This is the second seal.

**7.** You start as a baby learning to live life the proper way and live a harmony in all material, spiritual, mental, physical aspects of your life. Thus striving for MOKSHA. It is a constant daily struggle to live in purity, but at least this is the experiment. When I say you start as a baby, I mean you will go against what is normal and live by the word of scriptures. For example, everyone will be tempted to act as usual when someone insults you, or steals from you. Now the one who is reborn must respond to those things by what was written and not as they would normally do.

**202**

Being REBORN for the true believer or extreme serious person, is after taking Jesus as their savior to remove this Karma, that person changes their name. You can just go by another name even though your identification says your birth name. Or change your name legally. For calling you by your old name is useless for that person is dead. And don't talk about how great the old days of sin were: They are to be not spoken of and forgotten. Then slowly as a baby learning, leads a holy life according to scriptures. Buddha, Jesus dharma.

**8.** NOW THIS IS THE ENLIGHTENMENT! The previous version or person that was baptized DIED when going under the water. And was REBORN as a NEW PERSON when coming out of the water.

**9.** This is why ALL previous Karma from ALL previous lives is erased. That person doesn't exist anymore. Any attachment of any Karma or sins is on a dead person. YOU ARE NOT THAT DEAD PERSON ANYMORE. You are baptized with the Holy Ghost or Holy Spirit.

That is the experiment and if tested can Enlighten a great people from the subjection of never ending Karma. Allowing for this lifetime NOW to be the last and bring MOKSHA. So INDIA can keep their Gods. But for the Indians who want to

**203** end this cycle of continuous rebirths I am just adding one and the way. If it works follow Jesus fully. My intent only is to bring an end to those who wish to go home once and for all. If my Teleportation method is correct, then why not try this experiment. There is no hidden agenda. Just to bring India into a new era.

## Saty सत्य

How did you get to know your Karma was erased???
Sir you really think that Karma could be really erased by Baptism? As I said earlier there are many things in Hinduism that can erase our sins not Karma.

### J.V. जॉनी विं सेंटो

My friend Saty, you are a good Hindu, never straying past what is written, but I am going home after this life. No more rebirths. I will tell you what you asked in a long list of my payback Karma that was ERASED and stopped immediately after my ritual that I speak of through Jesus.

**1.** I was in back of police car and the prisoner next to me said "Why are we being arrested we didn't do anything these are false charges? Do you believe in Karma man?" I say "Yes, but the question is "HOW MUCH KARMA DO WE GOT COMING TO" just then, before I could finish the sentence a suicide driver came straight at us to ram the front of the police car. After we were avoiding that car on two wheels we were unharmed. The guy said "Do you think we would have been killed?" I replied "What do you think?

We are hand cuffed from behind with a steel cage in front of us?"
And before that incident everything bad was thrown at me one after another.

**2.** Dealership purposely ruined my car,

**3.** then got new car, that was ruined right away,

**4.** furnace went out when below zero,

**5.** false charges, jail police lawyers,

**6.** then more lawyers with someone else then arrested again,

**7.** got sick,

**8.** lost all my money,

**9.** then got robbed by mechanic,

**10.** then the other mechanic ruined my car,

**11.** then another ruined my tire.

**12.** Car vandalized,

**13.** passed out in jail a third time and broke my head open,

**14.** people vandalized my dad's truck and house thinking I lived there,

**15.** lawyer stole $600,

*reference #205*

**16.** threatened so much had fight.

**17.** judge was tired of seeing me wanted me to do a year in jail.

**18.** People putting nails in front of my residence,

**19.** broke my foot

**20.** on the 31st day after my 30 day warrenty my furnace went out again.

**21.** that all happened in a couple of years.

SO MY FRIEND, THAT IS A SUMMARY AND WHEN

JOHNNY VINCENTO

I GOT BAPTIZED ALL KARMA PAYBACK GONE. COMPLETELY STOPPED.

Note: When I was in the main cell, the gang leader of the block approached and asked me who I was. I said I was a preacher, (since I was so well versed in this subject.) The leader said "If you are a preacher then why am I in here?" I replied "What is your drug of choice?" He responded "Heroin." "You see my friend, the Lord saved you, because all over the news people are dying from that new heroin. People are putting Fentanyl in it, which is an animal tranquilizer right into the heroin." He started going around in circles, saying DAMN PREACHER THAT'S DEEP MAN!! I tell you what, This is our block, there is 18 of us. You're protected in here preacher."

I learned that when all the the material things are taken away and everyone is eating the same food, wearing the same clothes. EVERYONE IS EQUAL except for what is in their hearts. Another guy I met was in there for three months and he was the happiest person I ever met. I asked him how he could be so happy. He said "It is all about your mind. No one can take that away. They can lock my body up but not my mind."

But the evidence it was all karma is that there was no drivers on the road at about 3am. Just us locked in the back seat with hand cuffs. THE EVIDENCE IS WHEN I SAID **"THE QUESTION reference #206 IS HOW MUCH KARMA DO WE HAVE COMING TO"** That was done purposely by unseen forces to

try and kill me. And letting me know it was Karma and not a coincidence.
I may have been bombarded so much as a scientist, so I could find a way to remove Karma for the world, to understand through baptism of Jesus.

**Saty सत्य**
I got this sir. You have been misunderstood. Now I am not the Yogi or any master but these things you experienced are your sins as well as Karma of past life. Good Karma take over bad Karma and REDUCES it's effects but does not erase it. After that you baptized yourself even that was Karma. YOU DID NOT ERASE YOUR KARMA IN FACT I THINK YOU RAISED YOUR KARMA. YOUR TIME WAS PERFECT WHEN YOU GOT BAPTIZED BUT BAPTISM DIDNT ERASE YOUR KARMA IN FACT BAPTISM ITSELF IS A KARMA.
**207** REMEMBER ONE THING SIR KARMA IS THE ROOT AND SINS AND OTHER GOOD AND BAD THINGS ARE ITS NEVERENDING BRANCHES WHICH WE DO OUR ENTIRE REBIRTHS.

### J.V. जॉनी विं सेंटो
Saty, my assault of Karma totally stopped. It IS GONE and THAT IS A FACT. SO ANYONE WHO READS OUR GREAT CONVERSATION, MAKE YOUR OWN CHOICE: TEST IT. It is an experiment. It is worth trying to free billions of people in the future.

### Saty सत्य

Sir you have been misunderstood that your Karma is erased.

### J.V. जॉनी विं सेंटो

Jesus sent his disciple Saint Thomas to India to baptize the people and save them from the circle of Karma. That was two thousand years ago. It is an experiment. I don't have any proof, but what happened to me. Since I figured out how to TIme Travel and Teleport to overlapping worlds, I am a scientist people should listen to. This theory is not too extreme anyway.

Tell that to the old me laying on a jail floor with his head busted open and asking God "WHY HAVE YOU FORSAKEN ME?" Now I have peace. It is **208** wonderful, the pastor who put me under the water put his arm around me on the way out of the building later on,

I felt like I was walking with God Saty, it was the best day of all my life in this world.

### Saty सत्य

Sir I understand but you haven't erased your Karma. Karma is erased only when we get Moksha.

### J.V. जॉनी विं सेंटो

Yes, very true, the Baptisim of the ritual I described only cleans the slate to allow a person to strive for MOKSHA for the remainder of their life here on this Mortal plane of existence. Right after

Baptizim ritual Karma starts again. Live pure on all levels and one then CAN attain MOKSHA. The conclusion is not for us to decide. The people of India can decide. The churches of Saint Thomas and many other Christian churches are in many places in India right now for anyone who wants to test this theory. The name of Jesus's disciple was Thomas Didymus (Saint Thomas). Jesus always called Thomas his "Twin." In Greek Thomas means TWIN and in Hebrew Didymus means TWIN. Jesus's message was Thomas's. Thomas had the gifts of God Jesus.

**209**

### Saty सत्य

Sir, I am a true Hindu and you are a true Christian Guru. Do you really think that one simple ritual could change the entire people of India? Jesus is not in the Sanskrits?

### J.V. जॉनी विं सेंटो

Saty, I do know that the people can judge for themselves on if they want a way out to go home to paradise for good or stop terrible karma in this life.. I am a scientist, it is a valid experiment for anyone.

**Jesus is not in the Sanskrits, because he was a God who arrived after the Sanskrits to fix the issue of erasing all sins in all lives, all karma.** That is why he died. For that reason only. To show us that we can live forever by being reborn while still alive and finally go to Heaven. I guess very few were getting into Heaven.

**210**

### Saty सत्य

Please make sure you put this conversation in the book, the people will have both sides of our debate. I hope it does work, for you truely love the people of India. Raaja, Prisha, and my friends said the same thing. **211**

### J.V. जॉनी विं सेंटो

In conclusion, it is not good to have opinions as a researcher. This investigation has described where these manuscripts are. They need to be analyzed and dated to make sure they are not fakes. Or maybe as one said, After the book, the monks just made a manuscript. But monks are not deceivers. That is why they are monks.

I read many of Jesus scriptures, such as the Gospel of Thomas, Philip, Mary, Thomas the Contender and the Bible. Jesus has a real temper when it comes to people not listening to him. I see that in this work. Plus, this verse here:

11. "God will drive away the contaminated animals from His flocks; but will take to Himself those who strayed because they knew not the heavenly part within them."

I did see my father who died. I seen him in his quantum entangled physical body. I even handled his shoulder and arm. He was real alright.

That was in **212** February of 2022. He was an unbeliever, but yet there he was saved. He never knew he had any such heavenly part within him. Just the here and now and what could be seen and when you go that is it.

I don't think anyone could make up that verse.

In the lost book of Issa, there are very valid accurate arguments about the Hindu religion and their worship of a huge amount of good and evil gods along with idols. My Hindu student in India taught me his religion along with Buddhism for months. We wrote the book MY NAME IS SATY: THE RETURN OF TELEPORTATION.

Just that one book, shows how to turn India into the greatest spiritual and science superpower the world has ever seen. As a Christian, it was clear the solution for those two religions. To summarize quickly, there is over a billion people in India who have a religion that does not forgive sins. I even asked what type of religion is that? They have what is called "The Wheel of Samsara." It is an endless cycle of rebirths to Earth, due to having karma, to make up for all their previous lives sins.

Not to mention, any bad Karma they create in **213** this lifetime. Never being able to account for all their karma. So they are born again and again and again, a hundred times, a thousand times. I mentioned Jesus baptism for the solution. He died so all our sins could be forgiven and erased: In all previous lives. You the reader may not believe in previous lives, but there is a billion people who do.

GOT BAD KARMA? TEST IT!
JESUS / HOLY GHOST BAPTISM WILL ERASE ALL PREVIOUS LIVES SINS, STOPPING THE WHEEL OF SAMSARA SO YOU CAN WORK ON ACHIEVING MOKSHA AND FINALLY GO HOME TO PARADISE.

ONE'S BAPTISM IS MOST LIKELY THE MOST POWERFUL EXPERIENCE ANY PERSON CAN HAVE IN THIER LIVES. THE WEIGHT OF KARMA AND STACKED SIN BEING LIFTED OFF YOU IS LIFE CHANGING!

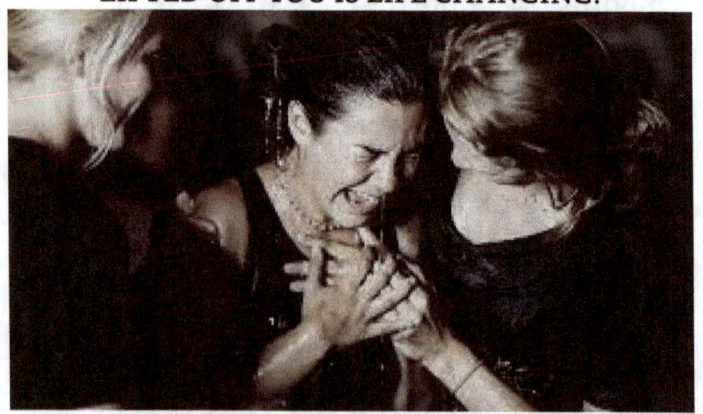

YOU MAY FEEL LIKE CRYING IN HAPPINESS JUST LOOKING AT THIS WOMAN'S OVERWHELMING JOY!!!

**Saty सत्य**
Guru, it is for all to see how you love India. It's great being your student, the Gods and the universes know we made a difference for the Satya Yuga. Your humble disciple of truth Saty.

## J.V. जॉनी विं सेंटो

The world is a bus station. Sending people to the worlds. The other worlds are already separated according to their frequencies of vibrational matter or degrees of darkness and light. Heaven is for the purest of people's inner LIGHT, the highest positive vibration of goodness.

Then you have the 2nd level world which is nice too, but those people's frequency is not pure enough for level one. After that plane of physical existence, you have the world of jerks and know it alls and condescenders who think the world revolves around them. That is level three.

At the lowest level is the famed and glorified world of Hell, which people laugh at and celebrate in movies and in real life, but that is where the bus is taking them and they don't even know it. They kill for fun and their function on Earth is to cause misery. But the only misery is their own in that world forever, for the bus drops them off and they get no rebirth, just everlasting misery in Hell. Jesus talks about that world in great detail.

So which bus are you going on? What ticket do you have? One's spirit leaves Earth and takes its particles to rematerialize an immortal body in an equal vibrational existence. It is impossible for a person with the lowest vibration to obtain the Resurrection into a higher vibration world. **Your ticket is your own frequency.**

This world is not separated, It is the transport station, with people all bunched together in a battle between LIGHT and darkness. Politics, the media and television is no longer politics or news or entertainment, but forces of good vs evil.

The bus drops off those who are going to Hell, but keeps bringing back the people who need to fix their karma back for another rebirth. An endless cycle of bus rides of dropping off in a new body and then trying to take you home. Over and over again for hundreds or thousands of times. But always Heaven says "Send that one back for another rebirth, that one's karma needs to be accounted for."

I am the messenger of the world. It doesn't matter who I am in prophecy. I am here to guide you home. Get Baptized and erase your karma to make this your last rebirth. Follow what is written here.

Make this your last trip to Earth. Have your

ticket aligned to go to Heaven or the 2nd level, which is nice too.

You can erase all present and previous lives karma, stopping the Wheel of Samsara.

Change your destination by activating the secret of the universe: Jesus baptism.

# FOR OTHER BOOKS BY THIS PHYSICIST GOOGLE JOHNNY VINCENTO

## WANT TO READ THE COMPLETE CONVERSATION? Google 'MY NAME IS SATY'

.99 cents Kindle Instant download worldwide - available in India

# TIBETAN MONKS ANTI GRAVITY METHOD

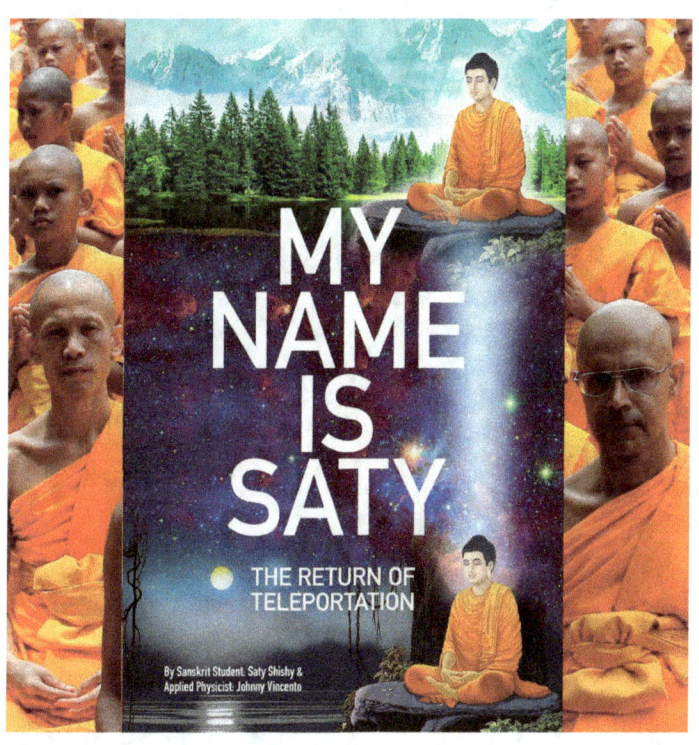

# HOW TO BECOME A MAJOR DUDE

# MOTIVATIONAL QUOTES FOR SCIENCE AND SPIRITUAL ENLIGHTENMENT

CAN BE USED EVERY YEAR - DAILY VERSES - LEAP YEAR QUOTE INCLUDED

## JOHNNY VINCENTO

DO YOU WANT A NEW LIFE OR A 360 CHANGE? Being a Major Dude is about APPLYING and understanding existence. My goal is to bring the entire planet into a new age of enlightenment. You can have your DREAM LIFE, your DREAM WIFE, your DREAM HOME with no mortgage. You may have never read my science. Nothing is impossible. I will show you that YOU WILL LIVE FOREVER. That is my science. No matter what you have been through, a person has the ability to actually achieve anything they want. All of your sins can be forgiven. You in fact, can become a NEW DIFFERENT human, BEING REBORN.

YOU CAN BECOME
PART MONK,
PART SCIENTIST,
PART JESUS CHRIST
A MAJOR DUDE
AND HAVE IT ALL!!!

Johnny Vincento is a physicist in the Applied field of full person Quantum Entanglement teleportation. He is one of the few people ever born to have seen Earth from outside its spacetime. He said "I am a scientist and a messenger to the entire world on what the TELEPORTATION PROJECT discovered. The physical world of Heaven itself."

# TIBETAN MONKS ANTI GRAVITY METHOD

Grab your coffee, you are going on a fantastic adventure. Taking you back to Jesus's time and putting together an amazing mystery that will have you feeling that you are solving history itself!

This ancient document is incredible, but could it be true? It accounts for Jesus's lost years from 12 to 30 years old and describes where his mortal body went (The resurrection happened, but people don't know about quantum entanglement) Most in the West do not know that Jesus's Disciple Thomas went to India. This is known as common knowledge in India and even the Presidents of that country speak highly of him in their speeches. They even made postage stamps to commemorate him. Eight of his chuches are still standing from around 2000 years ago.

The message of this Lost book will transform the entire world into one of LIGHT. The science is the second lost knowledge: Of how to bring in a NEW ERA, A NEW TRUTH: THE SATYA YUGA!!! ONCE HEAVEN IS KNOWN AS FACT, THEN PEACE WILL COME TO THE WHOLE EARTH: NO ONE WILL RISK BEING EVIL IF THEY WILL SPEND ETERNITY IN THE WORST PHYSICAL WORLD. PEOPLE'S FREQUENCIES JUDGE THEMSELVES.

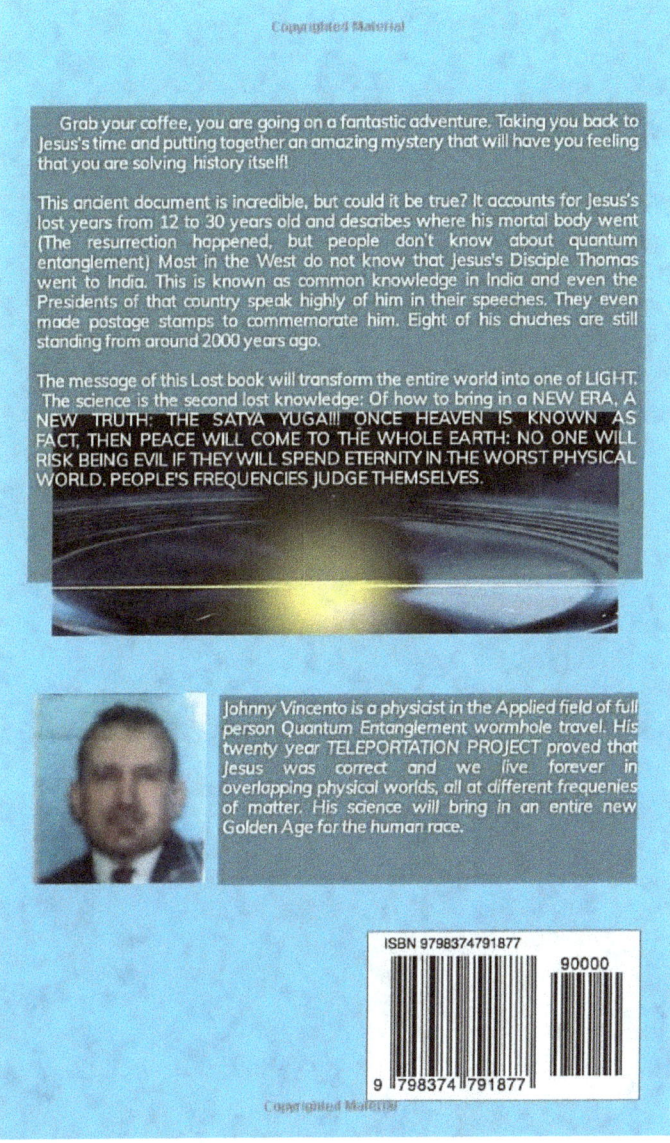

Johnny Vincento is a physicist in the Applied field of full person Quantum Entanglement wormhole travel. His twenty year TELEPORTATION PROJECT proved that Jesus was correct and we live forever in overlapping physical worlds, all at different frequenies of matter. His science will bring in an entire new Golden Age for the human race.

ISBN 9798374791877

*ATLANTIS ANCIENT TABLETS & THEIR MESSAGE TO HUMANITY ON SPACETIME TRAVEL*

TELEPORT NEWS

Johnny Vincento - Thoth - Plato

# TIBETAN MONKS ANTI GRAVITY METHOD

TELEPORT NEWS, brings you a rare paperback edition. Taking you right into lost eyewitness accounts of Atlantis. It describes the secrets of who built the Sphinx, Great pyramid and what they were built for. The science formula deciphered from these tablets has been experimented on for 20 years with great success. They were written ages ago in stone, estimated to be around 12,500 years old.

The writings were discovered and never understood since around 750 AD. They are written from an extremely advanced ancient scientist. As the world leading physicist in full person teleportation, I was able to decipher them. Only someone familar with these applied forces of quantum entanglement and wormhole travel to interdimensional parallel worlds, would understand what he is saying. I can see why no one understood them. The Project actually did exactly what was written and the findings are mind blowing. The ancient scientist said that he purposely hid the tablets for the future of mankind and what a story he has to tell us!!!

Johnny Vincento is a physicist in the Applied field of Quantum Entanglement Wormhole travel. He says "I'm not a pastor or a preacher, I am a scientist & messenger to the entire world on what the TELEPORTATION PROJECT discovered. I'm one of the few ever born to see Earth's exisitence from outside it's existence."

ISBN 9798355503369

# THOTH'S 12,500 YEAR OLD TREASURE MAP IN GIZA - IMMORTALITY THROUGH HIGHER LIGHT FREQUENCY

## TELEPORT NEWS

### Johnny Vincento

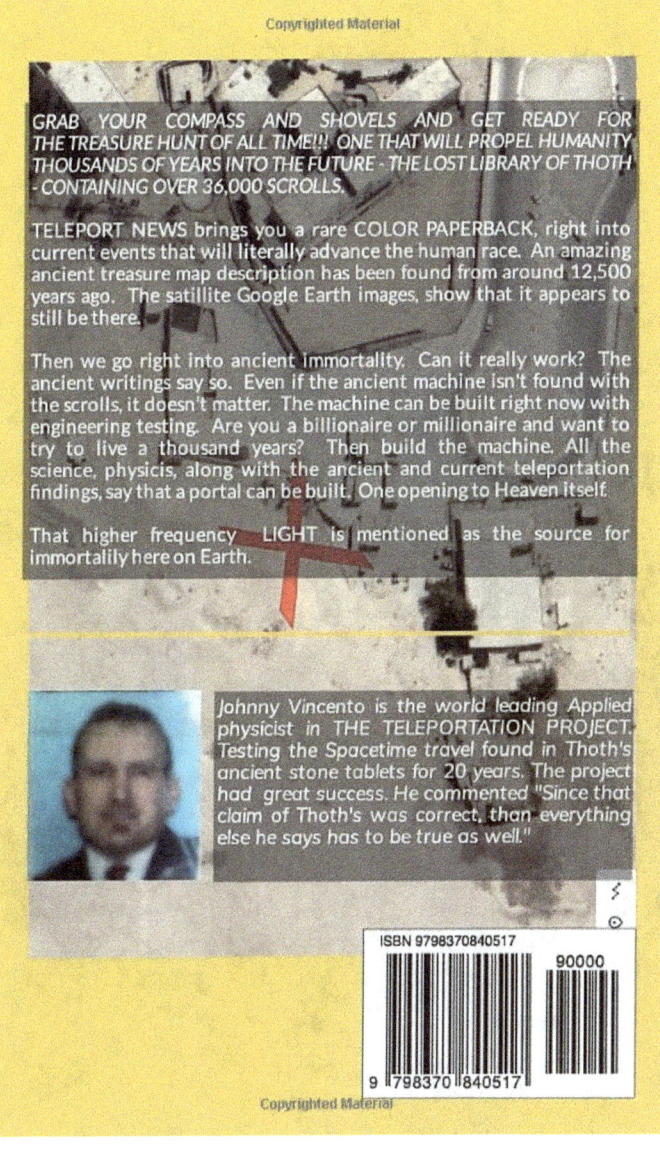

*GRAB YOUR COMPASS AND SHOVELS AND GET READY FOR THE TREASURE HUNT OF ALL TIME!!! ONE THAT WILL PROPEL HUMANITY THOUSANDS OF YEARS INTO THE FUTURE - THE LOST LIBRARY OF THOTH - CONTAINING OVER 36,000 SCROLLS.*

TELEPORT NEWS brings you a rare COLOR PAPERBACK, right into current events that will literally advance the human race. An amazing ancient treasure map description has been found from around 12,500 years ago. The satillite Google Earth images, show that it appears to still be there.

Then we go right into ancient Immortality. Can it really work? The ancient writings say so. Even if the ancient machine isn't found with the scrolls, it doesn't matter. The machine can be built right now with engineering testing. Are you a billionaire or millionaire and want to try to live a thousand years? Then build the machine. All the science, physicis, along with the ancient and current teleportation findings, say that a portal can be built. One opening to Heaven itself.

That higher frequency LIGHT is mentioned as the source for immortality here on Earth.

Johnny Vincento is the world leading Applied physicist in THE TELEPORTATION PROJECT. Testing the Spacetime travel found in Thoth's ancient stone tablets for 20 years. The project had great success. He commented "Since that claim of Thoth's was correct, than everything else he says has to be true as well."

ISBN 9798370840517

JOHNNY VINCENTO

# HOW TO CREATE FREE HYDROGEN VEHICLE FUEL FOR EUROPE

## WE ARE THE NEW WINNERS

Applied physicist Johnny Vincento

# TIBETAN MONKS ANTI GRAVITY METHOD

**FEATURED PRESENTATION IS IN FULL COLOR AT 8.5 X 11 INCHES**

THE FUTURE OF HUMANS IS AT A CROSSROAD. Obviously there are a lot of evil players and when all the cards are played on the Earth table, then humanity can go forward in a gigantic leap forward into enlightenment. Can you imagine a car that can travel a thousand miles on one fillup? Or 1500 miles? For practically free. These things are possible by just common knowledge. Well maybe not too common, but as the leading physicist, I explain it with toy scientists in their entire city called the "World Science Institute." These seemingly difficult science principles are illustrated in a clear way so all will be onboard. It is not that difficult to have a country as Sweden or the entire continent of Europe running on hydrogen free fuel. When I say free, I refer to hooking up your garden hose to a station in your driveway and creating free fuel. Not fantasy, for real. The system to set up throughout Europe seems absurdly easy. Each station is in fact independent from all others run on free energy to create the free fuel.

There are many other amazing discoveries that are ready to be REINTRODUCED into the world within this book. Well known by preflood humans, but long forgotton. You will be astounded by what has been refound.

Johnny Vincento is a physicist from Italy, in the Applied field of quantum entanglement wormhole teleportation. He said "In this book, I used toy scientists in their university to describe real truths of the universes and our hidden past. We are to be of LIGHT and the world leaders want to hide these truths. Heaven has been proven, we all live forever through quantum entanglement to the world that equals our own vibration. The story here can save entire countries if they listen to what these toy scientists are saying about our future: in just 6 years. Especially my beloved Sweden. All Swedish people need to get this book to save at least your country."

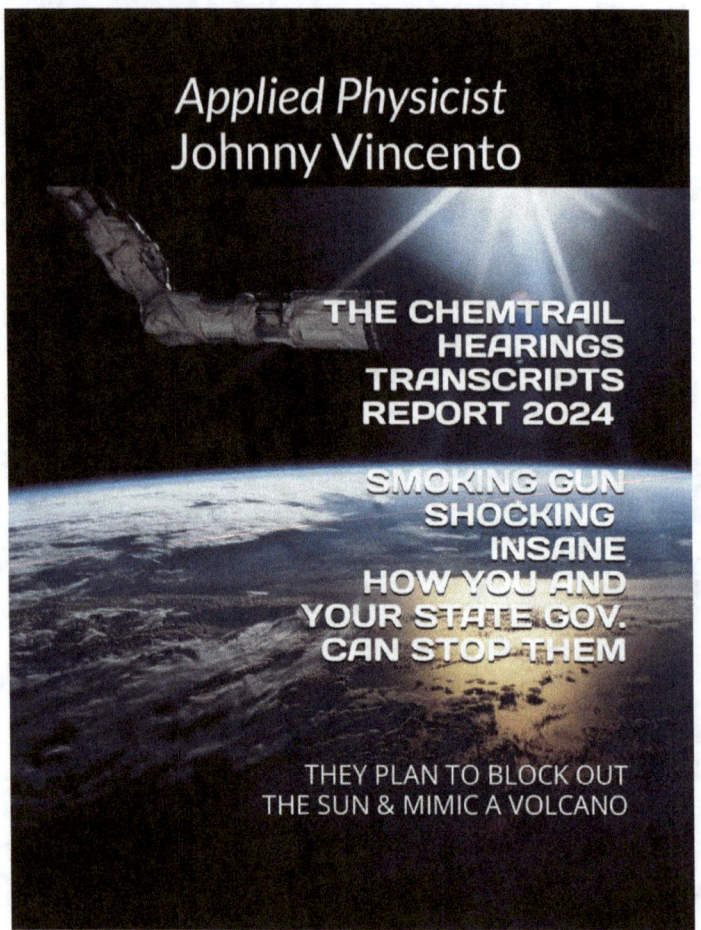

# TIBETAN MONKS ANTI GRAVITY METHOD

**The Chemtrail Hearings Transcripts**, show the U.S. government and its agents are commiting worldwide geoengineering disasters as attacks on the peoples of the world. Whether the excuse is not attacks, geoengineering to counter global warming, poison is still being sprayed on the populations and that is terrorism. You will see the exact chemicals and their effects on people, animals and the environment. The goal is a fraud and in reality is to increase global warming, sicken people and destroy food crops. Blocking out the Sun stops crops from growing, even if it cooled the Earth the crops would die.

The U.S. House of Rep. transcripts clearly plan other schemes such as blocking out the Sun in outer space and mimicking a volcano, all in the false claim of helping to cool the Earth. Even NASA said that these effects of the spraying trap heat in our atmosphere increasing global warming. A billionaire wants to buy millions of acres of land and cut down every tree and bury them: Disguised in the cause to counter global warming, but that INCREASES IT. Each state government and citizen now has the means to stop this terrorism in this REPORT 2024.

*Johnny Vincento is a physicist in the Applied field of Quantum Entanglement. He is known by many members of U.S. Congress, Senators, Governors, Government of Sweden and all Space agencies. He said "The only way humans will ever have peace is to see the other worlds that exist in front of them. All at different frequencies of vibrational matter. Invisible, but just as physical and real as Earth. As a radio tunes in to channel frequencies, so do the worlds, having invisible frequencies of matter. The Teleportation Project discovered we all live forever. Where do we go? To the world that equals our own particles vibration. For all these other worlds as Heaven - 2nd world (nice) - 3rd world (jerks) and Hell are just worlds of positive highest good vibrations - to negative lowest evil vibrations. Science and religion as one."*

# TIBETAN MONKS ANTI GRAVITY METHOD

Go back to 1852 and relive this exciting 2000 mile wagon train journey.

The 1852 diary of Esther Hanna is a day by day description of crossing the plains frontier. This true story is interlaced with other pioneer's diaries, some ahead, some behind. Many journals written the same days.

But when you hear "Circle the wagons, the Indians are coming!" What will you do?

JOHNNY VINCENTO

# GET THIS WORLD CHANGING BOOK. READY FOR ALL NATIONS TO TEST. Not fiction

## "THE TELEPORTATION PROJECT"

For your Home Library Display - *Special Promotion:*

*The Hardcover is same price as the Paperback.*

**ABOUT HARDCOVER AMAZON BOOKS:**
Beautiful Case laminate High Gloss Hardcover
Hardcover books are printed as case laminate. This means your hardcover book will not have a dust jacket and the art is printed directly on the cover.
This HARDCOVER BOOK is larger than the standard size. Book is 5.5 inches x 8.5 inches.
As one customer said "Very high quality book, it's a showpiece for my cabinet."

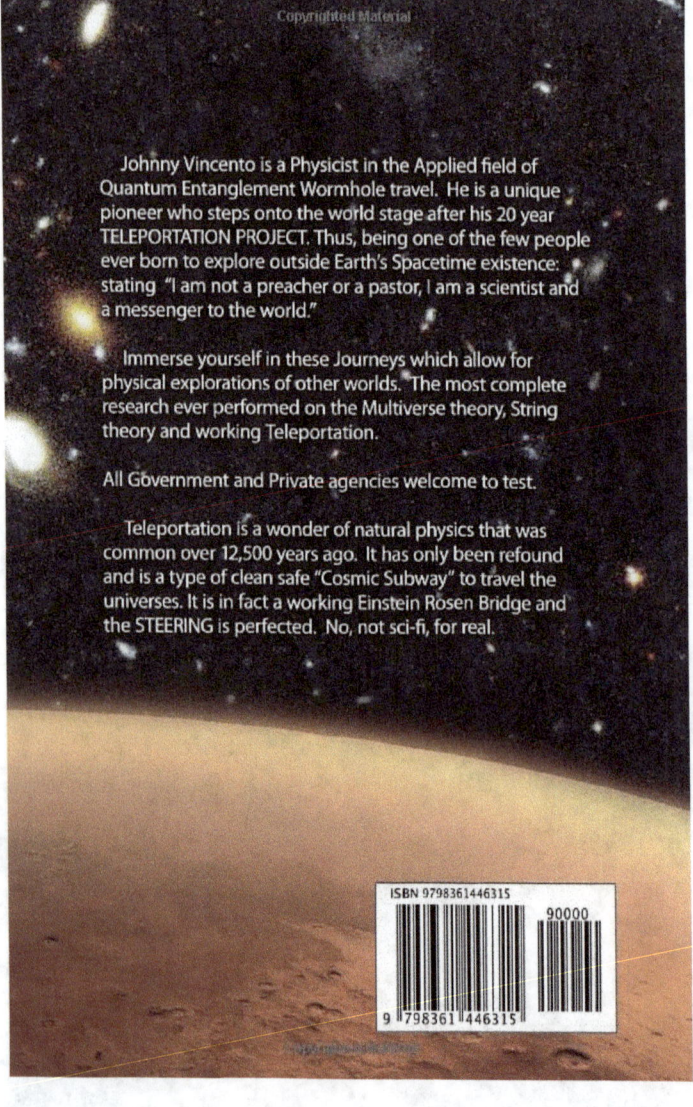

Johnny Vincento is a Physicist in the Applied field of Quantum Entanglement Wormhole travel. He is a unique pioneer who steps onto the world stage after his 20 year TELEPORTATION PROJECT. Thus, being one of the few people ever born to explore outside Earth's Spacetime existence: stating "I am not a preacher or a pastor, I am a scientist and a messenger to the world."

Immerse yourself in these Journeys which allow for physical explorations of other worlds. The most complete research ever performed on the Multiverse theory, String theory and working Teleportation.

All Government and Private agencies welcome to test.

Teleportation is a wonder of natural physics that was common over 12,500 years ago. It has only been refound and is a type of clean safe "Cosmic Subway" to travel the universes. It is in fact a working Einstein Rosen Bridge and the STEERING is perfected. No, not sci-fi, for real.

ISBN 9798361446315

*KINDLY LEAVE A REVIEW*

*Thank you*

www.ingramcontent.com/pod-product-compliance
Lightning Source LLC
Chambersburg PA
CBHW052257220526
45471CB00001B/383